III Conference la ValSe-Food and VI Symposium Chia-Link Network

III Conference la ValSe-Food and VI Symposium Chia-Link Network

Editors

Claudia Monika Haros
Loreto Muñoz

MDPI • Basel • Beijing • Wuhan • Barcelona • Belgrade • Manchester • Tokyo • Cluj • Tianjin

Editors
Claudia Monika Haros
Instituto de Agroquímica y Tecnología de Alimentos
(IATA-CSIC)
Spain

Loreto Muñoz
Universidad Central de Chile
Chile

Editorial Office
MDPI
St. Alban-Anlage 66
4052 Basel, Switzerland

This is a reprint of articles from the Proceedings published online in the open access journal *Biology and Life Sciences Forum* (ISSN 2673-9976) (available at: https://www.mdpi.com/2673-9976/8/1).

For citation purposes, cite each article independently as indicated on the article page online and as indicated below:

LastName, A.A.; LastName, B.B.; LastName, C.C. Article Title. *Journal Name* **Year**, *Volume Number*, Page Range.

ISBN 978-3-0365-5521-8 (Hbk)
ISBN 978-3-0365-5522-5 (PDF)

Cover image courtesy of Loreto Muñoz

Contents

About the Editors

Claudia Monika Haros

Claudia Monika Haros (Ph.D.), Chemist, graduated from School of Exact and Natural Sciences, University of Buenos Aires (UBA), Argentina, in 1990. She also has a M.Sc. in Bromatology and Food Technology (1992) and MSc in Biology Analysis (1997) from UBA. She has a Ph.D. in Chemistry (UBA, 1999), officially approved by the Ministry of Education, Culture and Sport from Spain. In 1991–2003, she worked as university professor in the Organic Chemistry Department, Food Science and Technology Area of UBA. In 1991–1999, she was Research Assistant in the Cereals and Oilseeds Group, Department of Industrial Chemistry, UBA. Later, in 2000–2002, she worked in Spain as a visiting professor in the Cereal Group of the Institute of Agrochemistry and Food Technology (IATA) in Valencia. In 2003, she was postdoc fellow at the Department of Food Microbiology, Institute of Animal Reproduction and Food Research (CENEXFOOD-EU), Polish Academy of Science, Olsztyn, Poland. In 2003–2004, she received an award for working with Prof. Sandberg of the Department of Chemical and Biological Engineering, Life Science Division, University of Chalmers, Gothenburg, Sweden. Since 2005, she has been a Research Associate (Ramon y Cajal Contract) of the Spanish Council for Scientific Research of the Ministry of Economy and Competitiveness (CSIC-MINECO). From 2008, she has been a permanent staff member of CSIC and continues her investigation in the Cereal Group, Department of Food Science of IATA. Between 2015 and 2021, she was the Coordinator of International Chia-Link Network and from 2018, she has been the Coordinator of la ValSe-Food Group-CYTED.

Since the early stages of her career, she has been mainly engaged in research with respect to Cereal Science and Technology field. The major theme in Dr. Haros's research is the utilization of different strategies to improve nutritional and/or functional value of cereal byproducts or cereal ingredients. In recent years, her research focused on nutritional studies of vegetable raw materials and/or their byproducts on their biological activity for their subsequent integration into new food matrices.

Loreto Muñoz

Loreto Muñoz (Ph.D.) Food Process Engineer from Universidad de Santiago de Chile (Chile), Master of Food Science from Universidad de Chile (Chile), Ph.D. in in Science and Food Engineering from Universidad de Santiago de Compostela (Spain) and Ph.D. in Engineering Science in Pontificia Universidad Catolica de Chile (Chile). At present, she is full professor at Universidad Central de Chile and Head of Food Science Lab (LabCial).

She has contributed to the training of advanced human capital through the direction of undergraduate and graduate thesis.

She has participated and directed numerous projects of national and international scientific and academic significance, as well as their impact through ISI scientific publications. In 2013, she received the BIMBO Panamericano Award for her research on the chia seed.

Currently, her research focuses on the extraction and characterization of food materials of plant origin of interest for human consumption, such as seeds, grains and legumes, in terms of their microstructural and morphological characterization, nutritional, physical (optical, rheological, textural and thermal properties), chemical and functional properties, as well as in the study of digestibility, bioaccessibility and bioavailability of nutrients. In addition, she has experience in extraction, separation and concentration of its components, as well as its application in food matrices with functional properties.

Finally, her current projects are related to the evaluation of new sources of dietary fibre in terms of the potential contribution on the reduction of risks associated with metabolic syndrome.

Preface to "III Conference la ValSe-Food and VI Symposium Chia-Link Network"

Iberoamerican crops are underutilized and cultivation levels are low, but, recently, their worldwide demand has significant increased, resulting in a production increase as well as a price increase. For many years, these valuable seeds have been widely recognized by food scientists and food producers because of their nutritional value. They contain high-quality proteins and some contain abundant amounts of starch and/or fibre (with unique characteristics) and large quantities of micronutrients such as minerals, vitamins and bioactive compounds; moreover, they are gluten-free, which makes them suitable for people suffering from gluten intolerances/allergies. For these reasons, the Iberoamerican valuable seeds interest has immensely increased, since in recent decades, research efforts have been intensified.

This book summarizes the Proceedings of the III International Ia ValSe-Food and VI Symposium of Chia-Kink Network "Iberoamerican Valuable Seeds for the Food of the Future", which was held in the Universidad Central de Chile, 15–17 October, Santiago, Chile. This book gathers the recent investigations of la ValSe-Food Group on these valuable seeds and other crops, and provides comprehensive and up-to-date knowledge of all the relevant fields of food science. It provides information on production and utilization, structure and chemical composition, paying special attention to carbohydrates, fibres, bioactive compounds, proteins and lipids of kernels and other parts of the plants. It includes their process, various food products and applications, and the nutritional and health implications. We hope that this book will contribute to the increased utilization of Iberoamerican valuable seeds in human nutrition.

This publication was financed by the Project Ia Valse Food-CYTED (Ref. 119RT0567).

Claudia Monika Haros and Loreto Muñoz
Editors

biology and life sciences forum

MDPI

Proceeding Paper

Nutritional Contribution of an Undervalued Ancestral Cucurbita, Study of *Sicana* sp. Endocarp, Epicarp and Seeds Composition †

Silvia Caballero, Eva Coronal, Anggie Burgos, Loida Galeano, Patricia Adelaida Piris Jara, Lourdes Wiszovaty, Cristian Oviedo and Laura Mereles *

Dirección de Investigaciones, Facultad de Ciencias Químicas, Universidad Nacional de Asunción, San Lorenzo P.O. Box 1055, Paraguay; scaballero@qui.una.py (S.C.); ecoronel@qui.una.py (E.C.); burgos_anggie@live.com (A.B.); loida21galeano@gmail.com (L.G.); ppiris@qui.una.py (P.A.P.J.); lourdesw@qui.una.py (L.W.); cristian_ovi_ro@hotmail.com (C.O.)

* Correspondence: laurameereles@qui.una.py

† Presented at the III Conference la ValSe-Food and VI Symposium Chia-Link Network, online, 15–17 November 2021.

Abstract: Native to South America, the *Sicana* sp. fruits, known in Paraguay as "kurugua", belongs to the Cucurbit family and is almost extinct in the region. The aim of this study was to determine the physicochemical characteristics, composition and antioxidant activity of "kurugua" with reddish peel color. The determinations were made by official and regional standardized methodologies on fresh weight (FW). The pulp has an alkaline pH (7.41 ± 0.11), and its main components are carbohydrates (9.44 ± 0.45 g·100 g^{-1}), followed by dietary fiber (1.74 ± 0, 04 g·100 g^{-1}), as minor proteins (0.53 ± 0.05 g·100 g^{-1}) and lipids (0.08 ± 0.01 g·100 g^{-1}). On the evaluated antioxidants compounds, they were higher in peel than in pulp as; total phenols (279.2 ± 12.1, 55.7 ± 10.3 mg of GAE·100 g^{-1}), Vitamin C (9.67 ± 0.09, 7.84 ± 1.71 mg·100 g^{-1}) and beta-carotene (0.37 ± 0.03, 0.19 ± 0.01 mg·100 g^{-1}), respectively. Fresh seeds have a high moisture content (38.8%), dietary fiber (40.2%) and lipids (11.74%), they mineral composition showed a high content of Mg and Ca and a high content of micronutrients such as Cu, Mn, Fe and Zn, which can represent a great contribution to the daily requirements of the diet. The red kurugua fruits are a natural source of nutritious and bioactive compounds beneficial to health, with multiple potential applications in foodstuff, which should be promoted in healthy dietary guidelines for the benefit of the populations.

Keywords: composition; carotenes; antioxidants; total phenol compounds; *Sicana* sp.

Citation: Caballero, S.; Coronal, E.; Burgos, A.; Galeano, L.; Jara, P.A.P.; Wiszovaty, L.; Oviedo, C.; Mereles, L. Nutritional Contribution of an Undervalued Ancestral Cucurbita, Study of *Sicana* sp. Endocarp, Epicarp and Seeds Composition. *Biol. Life Sci. Forum* **2021**, *8*, 1. https://doi.org/10.3390/blsf2021008001

Academic Editors: Loreto Muñoz and Claudia M. Haros

Published: 7 December 2021

1. Introduction

The kurugua fruit belongs to the genus *Sicana*, which in turn belongs to the cucurbitaceae family. This oval-shaped fruit, which grows on the vine similar to grapes, is native to South America [1]. Its pulp or endocarp is widely used by the inhabitants for various culinary recipes, including sweet and salty foods [2]. Despite having a very pleasant aroma and a tasty pulp, this fruit is not currently widely consumed in the region, due to lack of knowledge of its uses. This is because new generations and the urban population no longer grow them domestically. Unlike the fruits of *Sicana* sp. with reddish skin, the black *Sicana odorifera* variety is currently better known and studied [1,3]. The objective of this work was therefore to determine the physicochemical characteristics, proximal composition and antioxidant activity of the endocarp, epicarp and seeds of "kurugua" fruits *Sicana* sp. growing in Paraguay.

2. Materials and Methods

2.1. Plant Material

The fruits of *Sicana* spp. were collected in January 2020, from the department of Cordillera of the city of Juan de Mena (24°57′35.8″ S, 56°44′20.0″ W) Paraguay. The "kurugua" with the reddish peel color by convenience sampling of the harvest of the year 2020 in a mature state. They did not show any visible signs of damage and were sent to the Department of Food Biochemistry of the Faculty of Chemical Sciences of the National University of Asunción (FCQ-UNA). The seeds and epicarp were manually separated from the endocarp and analysed immediately.

2.2. Physicochemical Characteristics

Morphological studies were carried out on whole fruits without previous treatment, as described by Coronel et al. [3]. The pH (method N° 920.152), tithable acidity (method N° 925.53) and soluble solids (method N° 932.14) were measured according to AOAC Methods [4]. A potentiometer (BOECO, MBT-700 model, Berlin, Germany) at 25 °C and an analytical balance (KERN ADB, Baligen, Germany) were used.

2.3. Proximal Composition

The proximal composition of the endocarp and seed of the analysed fruits was determined by official methodologies (AOAC) [4]: moisture (method No. 950.06), ash (method No. 923.03), dietary fiber (method No. 991.42), total lipids (method No. 970.51) and total nitrogen using the conversion factor 6.25 from nitrogen to proteins (method No. 920.152). The content of total carbohydrates and soluble sugars was determined using the Clegg anthrone method, with and without previous acid hydrolysis, respectively [5]. The results were expressed in g/100 g fresh sample.

2.4. Minerals Content

The minerals content of the endocarp and seed of the analyzed fruits was determined by the atomic absorption (AA 6300 Shimadzu, Kyoto, Japan) AOAC method 975.03 [4]. The minerals analyzed were Na, Ca, Mg, Fe, Cu, Zn and Mn. For each mineral, a calibration curve was made using standard solutions (MERCK, Darmstadt, Alemania). The results were expressed in mg/100 g fresh sample.

2.5. Total Phenol Content

The content of total phenols was determined in the endocarp and epicarp of the fruits. The extracts were made with methanol: water (80:20), as described by IICA (2018) [6]. Total phenols were measured spectrophotometrically using the Folin–Ciocalteu reagent by the method described by Singleton and Rossi, (1965) [7], where the blue colored complex was quantified at 765 nm (UV-1800, Shimadzu, Kyoto, Japan). A gallic acid calibration curve (10–160 µg/mL) was used. The results were expressed in mg of gallic acid equivalents (GAE) per 100 g of sample fresh (mg of GAE/100 g).

2.6. Vitamin C Content

The Vitamin C content was determined in the endocarp and epicarp of the fruits. The determination was made using the spectrofluorometric method 967.22 of AOAC [4]. The results were expressed in mg of Vitamin C per 100 g of pulp fresh weight.

2.7. Content of β-Carotene

For the extraction of total carotenoids, the method previously described by Procisur, IICA was used [6]. The content of β-carotene was determined by HPLC-PDA with some modifications [6]. First, the extraction with BHT (in acetone) was performed. The injections were made immediately after each extraction. The chromatographic system used was: C18 column (Phenomenex Inc., Torrance, CA, USA) 250 cm × 4.6 mm, 5 µm, 100 Å, kept at 30 °C, FM: methanol: acetonitrile: triethylamine (900:100:1) isocratic. Flow 1.5 mL/min, injection volume 20 μL. Detector; PDA SPD-M20A (Shimadzu, Kyoto, Japan) at 450 nm. A calibration curve of β-carotene dissolved in HPLC grade acetone was used (0.3–3 μg/mL).

2.8. Statistical Analysis

The data were recorded in an Excel spreadsheet and analyzed in the statistical program Graphpad prism 5.0 (GraphPad Software Inc., San Diego, CA, USA). Student's t ($p \leq 0.05$) was used to determine the significant differences.

3. Results

3.1. Physicochemical Characteristics

The fruits of *Sicana* spp. analyzed has a reddish (Figure 1) peel that easily distinguishes it from the *Sicana odorífera* variety anthropurpurea, which has a black peel, as published in other works [1]. The morphological characteristics of the analyzed fruits are detailed in Table 1, where we can see that the weight of the fruits is quite similar to that reported in *S. odorífera* var. mature with black peel [1], as well as its transverse diameter; however, its longitudinal diameter turned out to be smaller. On the other hand, it was observed that the soluble solids and the pH of the endocarp of the kurugua with reddish skin analyzed are higher than the kurugua with black peel, however we found that the titratable acidity found in this work is lower than that reported by Coronel et al. [3].

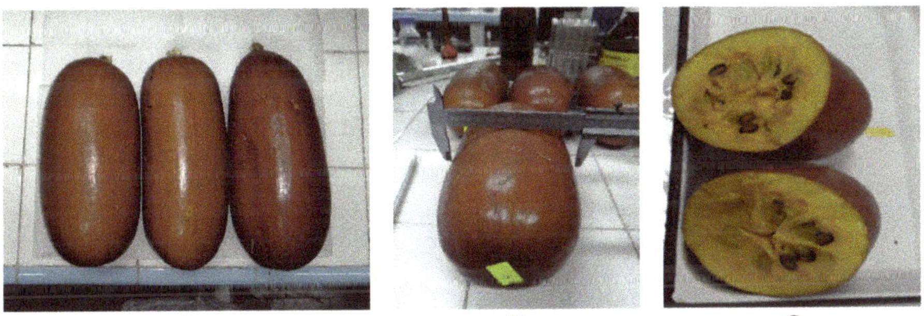

A B C

Figure 1. Reddish *Sicana* spp. fruits. (**A**) Whole fruits. (**B**) Measurements made. (**C**) Cross section of the fruit showing pulp and seeds.

Table 1. Physicochemical characterization of *Sicana* spp. with reddish peel.

Parameter	Result
Weight (g)	1656 ± 11
Longitudinal diameter (cm)	23.13 ± 1.82
Transverse diameter (cm)	10.90 ± 0.23
Soluble solids (°Brix) *	11.86 ± 0.21
Titrable acidity (g of Ac. Citrus) *	0.06 ± 10.02
pH *	7.41 ± 0.11

The values express the average of three repetitions ± SD.* Determinations made in fresh fruit endocarp.

3.2. Proximal Composition and Minerals Content of the Endocarp and Sed

The proximal composition and the mineral content of the endocarp and the fruit seeds analyzed are presented in Table 2, where we can observe that there are significant differences in all the determinations made (student's t, $p \leq 0.5$). This is to be expected due to it being a fresh fruit, where endocarp is also known as fruit pulp and has water as the majority component (86.67%), as reported by other authors [1,8]. In the seeds, the high content of dietary fiber (40.19%), total proteins (18.58%) and lipids (11.74%) stand out. Although the moisture content of the analyzed seeds (35.29%) is similar to that reported for the *Curcubita maxima* (pumpkin) [9], it has a much lower dietary fiber content and the total protein content is higher than that of this job. Of the minerals analyzed, Ca was the majority, both in the endocarp and in the analyzed seeds.

Table 2. Proximal and minerals content of endocarp and seed of *Sicana* spp. with reddish peel.

Parameter	Endocarp	Seed
Moisture (g/100 g)	86.67 ± 1.97 [a]	35.29 ± 0.04 [b]
Ash (g/100 g)	0.08 ± 0.01 [a]	2.64 ± 0.02 [b]
Total protein (g/100 g)	0.53 ± 0.05 [a]	18.58 ± 4.17 [b]
Total lipids	0.08 ± 0.01 [a]	9.16 ± 1.22 [b]
Total carbohydrate (g/100 g)	9.44 ± 0.45 [a]	3.35 ± 0.40 [b]
Dietary fiber (g/100 g)	1.74 ± 0.04 [a]	40.19 ± 0.00 [b]
Caloric Value (Kcal/100 g)	41	170
Na (mg/100 g)	3.46 ± 0.32 [a]	35.26 ± 0.02 [b]
Ca (mg/100 g)	21.21 ± 1.88 [a]	148.42 ± 3.74 [b]
Mg (mg/100 g)	1.58 ± 0.46 [a]	193.15 ± 4.84 [b]
Fe (mg/100 g)	0.25 ± 0.01 [a]	8.14 ± 0.63 [b]
Zn (mg/100 g)	0.42 ± 0.06 [a]	3.21 ± 0.01 [b]
Cu (mg/100 g)	0.19 ± 0.03 [a]	0.84 ± 0.05 [b]
Mn (mg/100 g)	0.42 ± 0.02 [a]	2.10 ± 0.05 [b]

The values express are means ± SD. Different letters indicate significant differences between means (*t* of Student $p \leq 0.05$).

3.3. Content of Total Phenols, Vitamin C and β-Carotene in the Endocarp and Epicarp

Content of total phenols, Vitamin C and β-carotene of the endocarp and epicarp of the fruits of *Sicana* spp. with reddish peel are shown in Figure 2. Significant differences (student's t, $p \leq 0.5$) are observed between the endocarp and epicarp in these three determinations, being the highest value for the epicarp in the case of phenol and β-carotene content and for the endocarp in the case of Vitamin C. The content of total phenols found in this work is higher than that reported in the pulp (endocarp) and peel (epicarp) of the kurugua with black peel [1], however, the Vitamin C content found in this work is less than the same [3]. The reddish color of the fruit's peel or the yellowish color could be due to a good content of β-carotene, although in this work it is shown that its content is low, so other substances that cause these colorations should be sought.

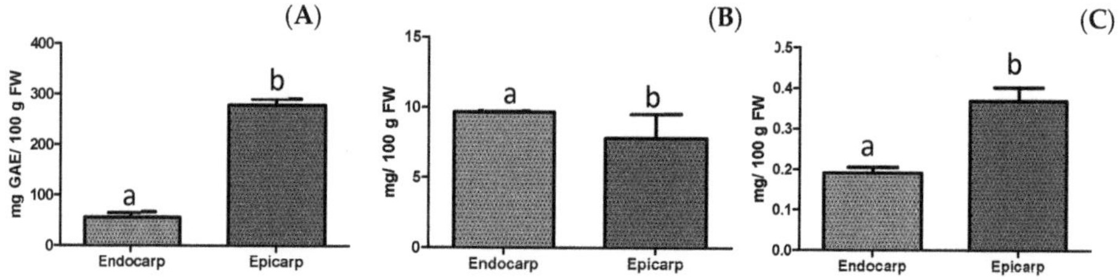

Figure 2. (**A**) Content of total phenols, (**B**) Vitamin C and (**C**) β-carotene of endocarp and epicarp of reddish *Sicana* spp. The values express are means ± SD. Different letters indicate significant differences between means (*t* of Student $p \leq 0.05$).

4. Conclusions

The morphological and physicochemical characteristics of the fruit of "kurugua" *Sicana* sp., and its proximal composition and minerals have been described, where the high content of dietary fiber, total proteins and lipids of the seeds, which are currently underused, stands out.

The presence of antioxidant compounds such as phenolic compounds and Vitamin C has been observed in the pulp and peel of the fruits. Significant amounts of carotenes have not been observed in the reddish rind or in the pulp of the fruit.

The fruits of *Sicana* sp. "kurugua" are a natural source of nutritional and bioactive compounds beneficial to health, with multiple potential applications in food, which should be promoted in healthy eating guidelines for the benefit of the population.

Institutional Review Board Statement: Not applicable.

Informed Consent Statement: Not applicable.

Data Availability Statement: Not applicable.

Acknowledgments: This work was supported by Ia ValSe-Food-CYTED (119RT0567). The authors are especially grateful to the "Kurugua poty" Foundation, Sr. Blas Imas, and Facultad de Ciencias Químicas-Universidad Nacional de Asunción for providing their facilities.

References

1. Mereles, L.; Caballero, S.; Burgos-Edwards, A.; Benítez, M.; Ferreira, D.; Coronel, E.; Ferreiro, O. Extraction of Total Anthocyanins from Sicana Odorifera Black Peel Fruits Growing in Paraguay for Food Applications. *Appl. Sci.* **2021**, *11*, 6026. [CrossRef]
2. Blanco, T. La Extraña Fruta Que Bautizó a Una Ciudad. *D. ABC Color* 2021, Español. Available online: https://www.abc.com.py/edicion-impresa/suplementos/abc-revista/2021/08/22/la-extrana-fruta-que-bautizo-a-una-ciudad (accessed on 15 September 2021).
3. Eva, C.; Silvia, C.; Rocio, B.; Rocio, V.; Laura, M. Sicana Odorifera "Kurugua" from Paraguay, Composition and Antioxidant Potential of Interest for the Food Industry. *Proceedings* **2020**, *53*, 10. [CrossRef]
4. Horwitz, W. *Official Methods of Analysis of the Association of Official Analytical Chemists*, 17th ed.; AOAC: Gaithersburg, MA, USA, 2000.
5. Dreywood, R. Qualitative Test for Carbohydrate Material. *Ind. Eng. Chem. Anal.* **1946**, *8*, 499. [CrossRef]
6. Instituto Interamericano de Cooperación para la Agricultura; Porgrama Cooperativo para el Desarrollo y Agroindustrial del Cono Sur. *Protocolos estandarizados para la valorización de frutos nativos del PROCISUR frente a la creciente demanda por ingredientes y aditivos especializados (carotenoides, antocianinas y polifenoles)*; Díaz, E.D., Pino, M.T., Saavedra, J., Eds.; PROCISUR: Montevideo, Uruguay, 2018; ISBN 978-92-9248-793-5.
7. Singleton, V.L.; Rossi, J.A. Colorimetry of Total Phenolics with Phosphomolybdic-Phosphotungstic Acid Reagents. *Am. J. Enol. Vitic.* **1965**, *16*, 144–158.
8. De Filho, G.X.P.; Barreira, T.F.; Pinheiro, S.S.; de Cardoso, L.M.; Martino, H.S.D.; Pinheiro-Sant'Ana, H.M. 'Melão Croá' (Sicana Sphaerica Vell.) and 'Maracujina' (Sicana Odorifera Naud.): Chemical Composition, Carotenoids, Vitamins and Minerals in Native Fruits from the Brazilian Atlantic Forest. *Fruits* **2015**, *70*, 341–349. [CrossRef]
9. Shahangir, A.H. Nutritional and Lipid Composition Analysis of Pumpkin Seed (Cucurbita Maxima Linn.). *J. Nutr. Food Sci.* **2015**, *5*, 374. [CrossRef]

biology and life sciences forum

MDPI

Proceeding Paper

Oil Characterization and Seeds Composition of *Sicana odorifera*, an Ancestral Cucurbita from Paraguay †

Laura Mereles *, Eva Coronel, Loida Galeano and Silvia Caballero

Dirección de Investigaciones, Facultad de Ciencias Químicas, Universidad Nacional de Asunción, San Lorenzo P.O. Box 1055, Paraguay; ecoronel@qui.una.py (E.C.); loida21galeano@gmail.com (L.G.); scaballero@qui.una.py (S.C.)

* Correspondence: lauramereles@qui.una.py

† Presented at the III Conference la ValSe-Food and VI Symposium Chia-Link Network, online, 15–17 November 2021.

Abstract: *Sicana odorifera* seeds, from an ancestral *Cucurbita* growing in Paraguay, possess important biowaste after fruit pulp use. However, there are reports that its infusions can reduce and cure the symptoms of viral diseases such as hepatitis, denoting its medicinal properties. The recovery of nutrients and bioactive molecules from its bio-residues has potential uses in the industrial sector with high added value as functional food ingredients. In *S. odorifera* species, although it is not a fruit for mass consumption, it is precisely the lack of a market for its biowaste that has limited its integral use. Based on this, the centesimal composition, oil characterization, and fatty acids profile of the kurugua seeds from two accessions (atropurpurea (black) and reddish) were studied. Kurugua seeds have been subjected to a cold extraction with a hydraulic press from dried whole seeds, and ISO and AOCS standard methods were used for analytical determinations. The major components in the centesimal composition of kurugua seeds were lipids, dietary fiber, and proteins. The oils presented iodine, saponification, and refractive indices characteristic of preferentially polyunsaturated oils. The major component in the fatty acid profile was linolenic acid, an important essential fatty acid in the diet. Although the characteristics of kurugua oil, demonstrate its potential application in the food industry as a polyunsaturated oil, source of essential fatty acids, future studies on stability and sensory analysis for food applications are suggested, with great possibilities for the food safety framework.

Keywords: biowaste; composition; Cucurbita; fatty acids; oil; *Sicana odorifera*

Citation: Mereles, L.; Coronel, E.; Galeano, L.; Caballero, S. Oil Characterization and Seeds Composition of *Sicana odorifera*, an Ancestral Cucurbita from Paraguay. *Biol. Life Sci. Forum* **2021**, *8*, 2. https://doi.org/10.3390/blsf2021008002

Academic Editors: Loreto Muñoz and Claudia M. Haros

Published: 2 December 2021

Publisher's Note: MDPI stays neutral with regard to jurisdictional claims in published maps and institutional affiliations.

1. Introduction

The ancestral species of the Cucurbitaceae family are part of the cultural and food heritage of several nations. However, some members of this family, such as *S. odorifera*, have been losing prominence to the point of being undervalued at the regional level, despite their delicate and delicious aroma and flavor and the multiple potential applications of its pulp [1].

The seeds of these fruits are important bio-residues. There are reports that its infusions can reduce and cure the symptoms of viral diseases such as hepatitis, denoting its medicinal properties. In *S. odorifera* species, although it is not a fruit for mass consumption, it is precisely the lack of a market for its biowaste that has limited its integral use. The recovery of nutrients and bioactive molecules from the bio-residues has potential uses in the industrial sector with high added value as functional food ingredients, which can be used by the health food industry [2].

The aim of this work was to characterize the composition of the seeds of *S. odorifera* in its two varieties (with fruits with black skin and reddish skin), which presents an opportunity to explore the use of biowaste from their pulp.

2. Materials and Methods

2.1. Plant Material

The fruits of *Sicana odorifera* collected in January 2020, from the Cordillera Department, Juan de Mena city (24°57′35.8″ S, 56°44′20.0″ W) Paraguay, are of two varieties; anthropurpurea (black) and reddish, harvested in 2020 in a mature state they did not show visible damage, and were sent to the laboratory for further analysis. The seeds were manually separated from the pulp and immediately analyzed to determine their physicochemical characteristics. For the centesimal composition, they were dried in a vacuum oven for 24 h at 60 °C.

2.2. Obtaining the Oil

The sample was weighed into a 250 mL Erlenmeyer flask and hexane was added maintaining the ratio 1:5 (p/v), this was stirred for approximately 3 h. Then, it was filtered under vacuum and subsequently the solvent was evaporated in a rotary evaporator (60 °C).

2.3. Analysis

The proximal composition analysis was determined by official methodologies [3] moisture (method No. 934.06), ash (method No. 968.08), dietary fiber (method No. 985.29), total lipids (method No. 948.22), and total nitrogen using the conversion factor 6.25 from nitrogen to proteins (method No. 970.39). The content of total carbohydrates and soluble sugars was determined by the Clegg anthrone method. The results were expressed in g/100 g on dry samples. The oil characterization parameters were measured by official AOCS methodologies (2009) [4], where the iodine index was performed by the Cd 1-25 method, the saponification index by the Cd 3-25 method and the refraction index by method Cc 7-25. The density corresponds to the magnitude that expresses the relationship between mass and volume. 1 mL of kurugua seed oil, black and red peel, was weighed, and later the mass-volume relationship was calculated. Kurugua seeds have been subjected to a cold extraction with a hydraulic press from dried whole seeds, and the ISO and AOCS standard methods were used for analytical determinations.

2.4. Statistical Analysis

The data were recorded in an Excel spreadsheet and analyzed in the statistical program Graphpad prism 5.0 (GraphPad Software Inc., San Diego, CA, USA). Student's t ($p \leq 0.05$) was used to determine the significant differences.

3. Results

3.1. Proximal Composition

The *S. odorifera* seeds of the analyzed varieties have similar shapes; they are flat oval and brown in color. Unlike the seeds of the black variety, which is uniform in color over the entire surface, the seeds of the red variety have a light brown interior and a darker brown halo on the edges. These characteristics are distinctive (Figure 1). In dry seeds, the major components in the centesimal composition of kurugua seeds were lipids (greater than 34%), dietary fiber (greater than 34%), and proteins (greater than 17%), as observed in Table 1.

(a) (b)

Figure 1. *S. odorifera* seeds from (**a**) black fruits and (**b**) reddish fruit.

Table 1. Centesimal composition and caloric value of *S. odorifera* kurugua seeds.

Parameter	Black Kurugua Seeds	Red Kurugua Seeds
Moisture (g/100 g)	10.63 ± 0.23 [a]	7.70 ± 0.26 [b]
Ash (g/100 g)	2.48 ± 0.14 [a]	2.64 ± 0.02 [b]
Total protein (g/100 g)	17.40 ± 0.81 [a]	18.55 ± 0.55 [a]
Total lipids	34.50 ± 0.28 [a]	36.3 ± 0.49 [a]
Total carbohydrates (g/100 g)	2.78±0.06 [a]	3.40 ± 0.32 [a]
Dietary fiber (g/100 g)	34.75 ± 0.27 [a]	39.94 ± 0.08 [a]
Caloric Value (Kcal/100 g)	391± 2.57 [a]	415± 4.67 [b]

The values expressed on dry weight, as the average of three repetitions ± DS. Different lowercase letters in each row indicate significant difference between the means (Student's t, $p \leq 0.05$).

3.2. Characterization of the Oil and Fatty Acid Profile of the Seeds

The Characterization of the oil from the seeds of *S. odorifera* are observed in the Table 2. On the other hand, the profile of the major fatty acids of the seeds oil are presented in Table 3. Eight different fatty acids were identified in the oil by GC/MS. Unsaturated fatty acids predominated in the seed oil with an average of 82.2%. The dominant fatty acid was C18: 3 omega 3 linolenic acid in both varieties (32.8–38.08%), together with C18: 2 omega 6 linoleic acid (28.62–29.52%) and acid oleic C18: 1 omega 9 (12.77–19.09%).

Table 2. Characterization of the oil from the seeds of *S. odorifera*.

Variety	Iodine Value (gI_2/100 g)	Saponification Index (mg KOH/g)	Density (g/mL), 25 °C	Refractive Index ND, 25 °C
Black fruit seeds	132.76 ± 2.29 [a]	166.88 ± 0.30 [a]	0.87 ± 0.02 [a]	1.479 ± 0.33 [a]
Reddish fruit seeds	130.58 ± 2.00 [a]	182.42 ± 0.00 [b]	1.09 ± 0.00 [b]	1.478 ± 0.00 [a]

Values are the mean ± SD of three determinations on oil seeds. Different lowercase letters in each column indicate significant difference between the means (Student's t, $p \leq 0.05$).

Table 3. Fatty acids profile of *S. odorifera* seeds oil.

Fatty Acids		Reddish Fruit Seeds (g/100 g Oil)	Black Fruit Seeds (g/100 g Oil)
Polyunsaturated	Linolenic (ω-3)	32.80 ± 0.13 [a]	38.08 ± 0.06 [b]
	Linoleic (ω-6)	28.62 ± 0.17 [a]	29.51 ± 0.21 [b]
	8.11 octadecadienoic Acid.	1.26 ± 0.03 [a]	1.18 ± 0.01 [a]
Monounsaturated	Oleic (ω-9)	19.39 ± 0.11 [a]	13.16 ± 0.55 [a]
	Palmitoleic (ω-7)	0.09 ± 0.01 [a]	0.06 ± 0.01 [a]
TOTAL		82.16	81.99
Saturated	Miristic (14:00)	0.12 ± 0.02 [a]	0.12 ± 0.01 [a]
	Palmitic (16:00)	10.98 ± 0.17 [a]	10.93 ± 0.08 [a]
	Estearic (18:00)	6.74 ± 0.07 [a]	6.96 ± 0.17 [a]
TOTAL		17.84	18.01

Values are the mean ± SD of three determinations on oil seeds.

4. Discussion

At a moisture level below 11%, the total lipid content in *S. odorifera* seeds observed is higher than that reported in seeds of fruits of the same family of Cucurbitaceae such as squash *Cucurbita maxima* (30.66 g/100 g) [5].

The oils presented iodine, saponification, and refractive indices characteristic of preferentially polyunsaturated oils. The iodine value indicates the degree of unsaturation of the fatty acids in an oil. According to these results, kurugua seeds would have the presence of unsaturated fatty acids, and classifies it as a "semi-drying oil". Our results for black and red kurugua seeds were similar to *Cucurbita moschata* Duch "zucchini or long-necked squash" (132.7 $gI_2/100$ g). However, the saponification index of the analyzed kurugua seeds (166.88 ± 0.30 and 182.42 ± 0.01 mg KOH/g, in black and red *Sicana sp.* seed oils, respectively), was higher than that reported for *C. moschata* (122.90 mg KOH/g) [6].

The fatty acid composition of the seeds depends on their genetic characteristics, but the latitude and climatic conditions of cultivation also have a strong influence on fatty acid biosynthesis. The observed values of the content of these fatty acids are lower than those reported for chia seeds, recognized as a source of essential fatty acids [7]. Although the oil from kurugua seeds has not been previously characterized in the light of our knowledge, the seeds of *S. odorifera* that grow in Paraguay can provide essential fatty acids such as omega 3 and omega 6. These results demonstrate their nutritional qualities, with potential for use as a food ingredient in healthy prepared food preparations.

5. Conclusions

S. odorifera seeds have a lipid content with good nutritional characteristics, preferably polyunsaturated fatty acids, were omega 3 linolenic acid was the mayority. Future studies on stability and sensory analysis for food applications are suggested, with potential aplications, on food safety framework.

Institutional Review Board Statement: Not applicable.

Informed Consent Statement: Not applicable.

Data Availability Statement: Not applicable.

Acknowledgments: This work was supported by grant Ia ValSe-Food-CYTED (119RT0567), "Kurugua poty" Foundation and Canindeyú Department Government. Bioq. Allison Alegre and the student Elaine Ibarrola, for their collaboration in the analysis of the fatty acid profile.

References

1. De Paula Filho, G.X.; Barreira, T.F.; Pinheiro, S.S.; de Morais Cardoso, L.; Duarte Martino, H.S.; Pinheiro-Sant'Ana, H.M. 'Melão croá' (Sicana sphaerica Vell.) and 'maracujina' (Sicana odorifera Naud.): Chemical composition, carotenoids, vitamins and minerals in native fruits from the Brazilian Atlantic forest. *Fruits* **2015**, *70*, 341–349. [CrossRef]
2. Albuquerque, B.R.; Dias, M.I.; Pereira, C.; Petrović, J.; Soković, M.; Calhelha, R.C.; Oliveira, M.B.P.P.; Ferreira, I.C.F.R.; Barros, L. Valorization of *Sicana odorifera* (Vell.) Naudin Epicarp as a Source of Bioactive Compounds: Chemical Characterization and Evaluation of Its Bioactive Properties. *Foods* **2021**, *10*, 700. [CrossRef] [PubMed]
3. Horwitz, W. *Official Methods of Analysis of the Association of Official Analytical Chemists*, 17th ed.; AOAC: Gaithersburg, MA, USA, 2000.
4. Firestone, D. (Ed.) *Official Methods and Recommended Practices of the American Oil Chemist's Society AOCS*, 6th ed.; AOCS: Urbana, IL, USA, 2009.
5. Shahangir, A.H. Nutritional and Lipid Composition Analysis of Pumpkin Seed (*Cucurbita maxima* Linn.). *J. Nutr. Food Sci.* **2015**, *5*, 374. [CrossRef]
6. Ortiz Grisales, S.; Pasos López, S.; Abadía Rivas, X.; Valdés Restrepo, M.P.; Vallejo Cabrera, F. Extracción y caracterización de aceite de semillas de zapallo. *Acta Agronómica* **2009**, *58*, 145–151. Available online: https://www.redalyc.org/articulo.oa?id=169916219005 (accessed on 15 September 2021).
7. Hernández-Pérez, T.; Valverde, M.E.; Orona-Tamayo, D.; Paredes-Lopez, O. Chia (*Salvia hispanica*): Nutraceutical Properties and Therapeutic Applications. *Proceedings* **2020**, *53*, 17. [CrossRef]

biology and life sciences forum

MDPI

Proceeding Paper

Sustainable and Circular Food Innovation—The *CeTA* Experience †

Juan P. Vivanco

Centro Tecnológico para la Innovación Alimentaria (CeTA). Ruta 68 N° 18.501, Reserva Cora N° 6, Pudahuel, Santiago de Chile 9020000, Chile; jpvivanco@cetalimentos.cl

† Presented at the III Conference Ia ValSe-Food and VI Symposium Chia-Link Network, online, 15–17 November 2021.

Abstract: Worldwide, around a third of loss and waste is generated at different stages of the food transformation chain, generating relevant economic, social, and environmental impacts, and increases in the water footprint, emission of greenhouse gases, pressure on the use of arable land, production costs, and decrease in the availability of food for the population. These reasons make imperative the implementation of strategies that minimize the generation of these losses. The Chilean "Technology Center for Food Innovation" (*Centro Tecnológico para la Innovación Alimentaria*—CeTA), aware of this problem, is contributing to the development of innovative products where materials that are considered waste or by-products from processes in the food, agriculture, cattle raising, and aquaculture industry are reused, or raw materials that do not meet commercial standards, taking advantage of their properties and bioactive compounds, turning them into value propositions that have circular economy components. Examples of these products developed in CeTA include soups, fruit purees, snacks, baked products, food ingredients, and breakfast cereals that contain valued raw materials such as barley bagasse, defatted coconut flour, fruit pomaces, discarded meats, quinoa grown in lagging areas of Chile, as well as stems, leaves, and fruit and vegetable peels, thus generating an environmental, economic, and social impact.

Keywords: byproducts; circular economy; food innovation; sustainability; upcycling; wastes

Citation: Vivanco, J.P. Sustainable and Circular Food Innovation—The *CeTA* Experience. *Biol. Life Sci. Forum* **2021**, *8*, 3. https://doi.org/10.3390/blsf2021008003

Academic Editors: Loreto Muñoz and Claudia M. Haros

Published: 15 December 2021

Publisher's Note: MDPI stays neutral with regard to jurisdictional claims in published maps and institutional affiliations.

1. Introduction

Food loss and waste are some of the main world problems. According to FAO, 1/3 of the food produced for human consumption is lost or wasted worldwide, that is, about MM 1300 tons of food per year [1]. This generates a decrease in the availability of food for the population, loss of resources used in its generation, increased production and final costs, causing unnecessary CO_2 emissions (3.3 million tons CO_2 and greenhouse gas emissions 6–10%), wasting water resources (250 km^3 blue footprint), increasing pressure on arable land, among others [2]. In Chile, for fruit and vegetable industries, 52% of the country's fruits and vegetables go to processing in 246 industries, which would produce 4.6 million tons of waste per year from this raw material [3]. On the other hand, there is a high level of undernourishment due to the tons wasted by Latin America (127 million tons/year) capable of meeting the nutritional requirements of 300 million people [4]. In Chile, the national strategic program *Transforma Alimentos* carried out a characterization of six agri-food chains of importance for the country (olive oil, apples, blueberries, plums, cherries, and tomatoes) with the aim of investigating their current and potential uses and proposing improvements to minimize losses and value generated by-products [5]. Recently, the Government of Chile launched the circular economy roadmap, whose vision is that by 2040, the regenerative circular economy will drive Chile towards a sustainable, fair, and participatory development that puts people's well-being at the center. This, through the care of nature and its living beings, the responsible and efficient management of natural resources, and a society that uses, consumes, and produces in a sustainable and conscious way, promotes the creation of green jobs and opportunities for people and organizations throughout the country [6].

The Chilean "Technology Centre for Food Innovation" is a public-private, non-profit corporation whose mission is for Chile to become a global player in the development and production of sophisticated and sustainable foods, in line with the circular economy global trend. To accomplish this, it has three innovation centers, located in strategic areas of Chile, forming and consolidating a national network of pilot plants, an infrastructure that will allow prototyping, piloting, and scaling food innovations, adding value to raw materials and agri-food by-products.

2. Results and Discussion

A list of the innovation projects focused on circular economy and quinoa valorization that have been developed by Chilean entrepreneurs, start-ups, private companies, universities, and non-profit organizations in collaboration with CeTA, along with a brief description of each of them is shown below:

2.1. Development of Healthy Breakfast Cereal Flakes from Quinoa of Aymara Origin (CORFO InnovaChile, Code 20SN-129433)

This innovation project consists of addressing the problem of the low commercialization of Chilean quinoa by indigenous communities. This initiative was carried out by *Suma Juira* indigenous association, located in the northern Chilean village of Cariquima (lagging zone from Colchane district, Tarapacá Region), which has 33 founding partners belonging to nine Aymara indigenous communities. This project's aim is the development of healthy quinoa-based flakes breakfast cereal with high nutritional added value: No sugar added, without warning labels for critical nutrients associated with Chilean law of food labeling and advertising, impacting this indigenous community in both economic and social terms, through a value proposition for a valuable ancient seed.

2.2. Valorization of Food Surpluses for the Formulation of Solidary and Healthy Foods in Wholesale Markets (FIC-RM, Code 40026935-0)

The sale of fruits and vegetables in *Lo Valledor* wholesale market—located in Santiago, Chile—generated around 18 thousand tons of organic waste in 2018. As a result of the recommendations of a study carried out by the *Universidad Bernardo O'Higgins* (UBO), the *Banco de Alimentos Lo Valledor* foundation (FBALV) was founded which has the aim to recover fruits and vegetables in order to give them free of charge to social organizations that provide food to vulnerable people. In the search for new solutions for this matter, a circular economy strategy in the *Lo Valledor* wholesale market is being implemented by these two entities (UBO and FBALV) in collaboration with CeTA, funded by the Government of Chilean Metropolitan Region consisting in valuing surpluses (leaves and stems discarded from the sale of vegetables and fruits, and fruits and vegetables not used by FBALV) in the preparation of sustainable and healthy foods: Soups, fruit preserves, concentrates. In this way, innovation will be promoted for both satisfying the need to expand food upcycling by means of valorizing food surpluses and the need of the social organizations belonging to FBALV to diversify their food matrix, making available healthy foods for free focused on the needs of beneficiaries.

2.3. Valorization of Barley Bagasse in the Development of Breakfast Cereals with Healthy Properties (CORFO InnovaChile, Code 20SN-151694)

The beer market in Chile has grown robustly in recent years. Barley bagasse is a by-product that corresponds to about 25% of beer production: This by-product represents an opportunity to reinsert a renewable input into a production process due to its remarkable content of dietary fiber and protein, with a positive environmental impact. Making the most of it, *Triunfo* brewery is developing a breakfast cereal with the addition of barley bagasse and healthy properties (high content of dietary fiber, non-sugar added, without warning labels for critical nutrients associated with Chilean law of food labeling and advertising) and a circular economy component. This innovation project is contemplating the prototyping, productive scale-up, nutritional, and sensory validation of this breakfast cereal, constituting an important value proposition towards sustainable beer production.

2.4. Reuse of Grape Pomace for the Development of Healthy Extruded Snacks (CORFO InnovaChile, Code 21SEC-171242)

The wine industry is of great importance in the central region of Chile. Each year, it generates a lot of waste, such as grape pomace and seeds, with significant potential to be revalued due to its important content of dietary fiber and polyphenolic compounds. This innovation project focused on circular economy carried out by *B-Japi* entrepreneur, attends to develop a dippable healthy snack by means of the reuse of red grape pomace as a source of sustainable raw material that allows obtaining foods with healthy properties at lower costs. Thanks to the use of the microwave vacuum dehydration technology and subsequent cook extrusion, it is possible to develop a food that preserves nutritional, chemical, and bioactive properties present in the grape pomace from red wine, which may contribute to prolonged satiety, good digestive function of the intestinal tract, and protection of the bacterial flora.

2.5. Keto-Friendly and Protein Pop Dry Snacks: Remanufacturing and Rethinking the Meat Industry (CORFO InnovaChile, Code 21SEC-171153)

In general terms, the waste from slaughterhouses and other meat processing industries is used in the production of animal feed or simply discarded, which is associated with environmental problems that should be avoided. In the framework of the circular economy, considering its environmental, economic, and social aspects, this innovation project carried out by *Llawken* company will give rise to a new value proposal, based on the use of these by-products from the meat industry, creating snacks low in available carbohydrates (keto-friendly) using an emerging and sustainable technology, such as vacuum microwave-assisted dehydration (also knowing as "pop drying"), which has the advantages of maximizing the nutritional quality of the product and its sensory characteristics. The development of a recyclable packaging that has oxygen, light, and moisture barriers will also be carried out.

2.6. Upcycling in the Fruit and Vegetable Industry: Healthy Snacks Based on Discarded Vegetables (CORFO InnovaChile, Code 21SN-182583)

"The Imperfect Project" Chilean start-up is an initiative that seeks to generate reuse (upcycling) of fruits and vegetables that do not meet some commercial quality standards required to sell it as is and/or its agri-food by-products, in the development of new food products for massive consumers, such as snacks and cereal bars, maintaining their healthy properties. This would make it possible to face problems associated with overweight and obesity in the population caused by processed foods, taking into account that post-COVID19 data show a growth in world consumption of this kind of products. It is an innovation project with environmental, social, and commercial impact, focused on a powerful and growing snack market.

2.7. Healthy and Sustainable Extruded Snacks Based on Local Legumes, Plants, and Fruits (CORFO InnovaChile, Code 21SNM-156926)

The robust growth in the snack market category in Chile allows the generation of opportunities for *Verax*—a women´s entrepreneurship—to develop new food products in this category with a focus on a healthier diet in comparison to most of the snacks available in the market, using local raw materials such as fruits, vegetables, legumes, and apple fiber. The latter is a by-product obtained from the processing of this fruit, constituting an important proposal for the revaluation of resources with a positive environmental impact. That is why the final objective of this innovation project is to develop a portfolio of extruded snacks free of warning labels for critical nutrients associated with Chilean law of food labeling and advertising, with the addition of pre and probiotics, with healthier properties for the immune and digestive systems than conventional snacks, and a circular economy component. The project focuses on prototyping of the products and packaging, scale-up, as well as nutritional and sensory validation.

2.8. Valorization of Defatted Coconut Flour By-Product to Develop Healthy Breakfast Cereals (CORFO InnovaChile, Code 21SN-182567)

The increase in awareness in relation to health and integral well-being has generated more informed consumers who are interested in eating good, nutritional, and healthy quality foods. Due to this new trend, the sale of products such as coconut fat—used for both culinary and aesthetic purposes—has increased, generating a great quantity of by-products associated with its production. This innovation project is carried out by a Chilean company *Nutrisa*, under the frame on the need to generate a circular economy-focused product by using the defatted coconut by-product flour (rich in dietary fiber) in an optimal, innovative, and beneficial way, giving it added value by incorporating it as an ingredient with high nutritional properties to develop a healthy breakfast cereal with differentiating and striking attributes for current consumers such as two products obtained through the cook extrusion process, one with antioxidant properties and the other with probiotics, products to which novel flavoring systems will also be applied.

3. Conclusions

The food development and innovation initiatives supported by CeTA contemplate co-creative work with the entrepreneurs and beneficiaries of the projects from the idea detection based on a problem or opportunity, diagnosis and review of the state of the art, creation of the business model, to then execute the development stages of the product(s) through ideation stages, concept product tests, application and small-scale prototype execution, scale-up at a pilot plant level, safety, nutritional and sensory issues, in addition to being profitable value propositions that contemplate environmental impacts with circular economy components and upcycling of food by-products and waste.

Institutional Review Board Statement: Not applicable.

Informed Consent Statement: Not applicable.

Acknowledgments: Author thanks Ia ValSe-Food-CYTED (119RT0567) grant for the invitation to present this work.

References

1. Gustavsson, J.; Cederberg, C.; Sonesson, U.; van Otterdijk, R.; Meybeck, A. *Pérdidas y Desperdicio de Alimentos en el Mundo—Alcance, Causas y Prevención*; Organización de las Naciones Unidas para la Alimentación y la Agricultura (FAO): Roma, Italia, 2012.
2. HLPE. *Las Pérdidas y el Desperdicio de Alimentos en el Contexto de Sistemas Alimentarios Sostenibles*; Un informe del Grupo de Alto Nivel de Expertos en Seguridad Alimentaria y Nutrición del Comité de Seguridad Alimentaria Mundial: Roma, Italia, 2014.
3. Eguillor-Recabarren, P.; Acuña-Reyes, D. *Pérdida y Desperdicio de Alimentos (PDA) en Chile: Avances y Desafíos*; Oficina de Estudios y Políticas Agrarias (ODEPA), Ministerio de Agricultura: Santiago, Chile, 2019.
4. Giraldo de los Ríos, A.C.; Bañados-Torres, N.; Leiva-Ramírez, F. *Guía para Prevenir y Reducir la Pérdida de Frutas y Hortalizas*; Organización de las Naciones Unidas para la Alimentación y la Agricultura (FAO) y Oficina de Estudios y Políticas Agrarias (ODEPA), Ministerio de Agricultura: Santiago, Chile, 2019.
5. Rossier, F.; de Feria, F.; Sepúlveda, A. *Cero Pérdida de Materia Prima en la Agroindustria—Informe de Etapa 1: Diagnóstico*; Transforma Alimentos: Santiago, Chile, 2018.
6. MMA. *Hoja de Ruta para un Chile Circular al 2040*; Ministerio del Medio Ambiente (MMA): Santiago, Chile, 2021.

biology and life sciences forum

MDPI

Proceeding Paper

Physicochemical Properties of *Moringa oleifera* Leaves Grown in Valencian Community (Spain) †

María D. Ortolá [1,*], María Luisa Castelló [1], Maria C. Etchevers [1], Francisco José García-Mares [2] and María D. Soriano [3]

1 Institute of Food Engineering for Development, Universitat Politècnica de València, Camino de Vera s/n, 46022 Valencia, Spain; mcasgo@upv.es (M.L.C.); coni.etchevers@live.cl (M.C.E.)
2 Department of Hydraulic Engineering and Environment, Universitat Politècnica de València, Camino de Vera s/n, 46022 Valencia, Spain; fjgarcia@gmf.upv.es
3 Department of Plant Production, Universitat Politècnica de València. Camino de Vera s/n, 46022 Valencia, Spain; asoriano@prv.upv.es
* Correspondence: mdortola@tal.upv.es

† Presented at the III Conference la ValSe-Food and VI Symposium Chia-Link Network, online, 15–17 November 2021.

Abstract: *Moringa oleifera* is a foliated tree widely cultivated in tropical latitudes, which is highly adaptable to climatic conditions and dry soils. Every part of the plant has nutritional, therapeutic or industrial benefits. This is due to its phytochemicals such as glucosinolates, phenolic compounds, alkaloids, terpenoids and tannins, high values of crude protein, carbohydrates, starch and lipids. In addition, the use of the leaves has increased considerably by the agro-food and biochemical industries since they are a valuable source of dietary proteins and essential amino acids. This work aimed to characterize three types of leaf from *Moringa oleifera* seeds with different origins (Thai (C1), Ghana (C2) and India (C3)), grown in the same plot, but with different cultural practices (intended for leaf production (C1 and C2) or sheath production (C3). For this, water content and optical properties were determined in the fresh leaves. Later the leaves were dried (50 °C for 8 h) and pulverized, analyzing their water content, antioxidant capacity, color and amino acid content. No significant differences were observed in fresh leaves in terms of humidity and color. In dry powder, a higher antioxidant capacity was registered in moringa type C2, with a% DPPH inhibition of 83.7%, although in all cases, it exceeded 60%, showing the high persistence of the antioxidants after drying. Serine, glutamic acid and alanine were the major amino acids with values of 373 ± 78, 301 ± 51 and 248 ± 9 mg/100 g of powder, respectively, without influencing the applied field treatment or origin.

Keywords: amino acid content; antioxidant capacity; color; *Moringa oleifera*

Citation: Ortolá, M.D.; Castelló, M.L.; Etchevers, M.C.; García-Mares, F.J.; Soriano, M.D. Physicochemical Properties of *Moringa oleifera* Leaves Grown in Valencian Community (Spain). *Biol. Life Sci. Forum* **2021**, *8*, 4. https://doi.org/10.3390/blsf2021008004

Academic Editors: Loreto Muñoz and Claudia M. Haros

Published: 15 December 2021

1. Introduction

The opportunity of diversifications offered by the agri-food sector implies the need to address the development of certain crops due to the new climatic conditions. The increasing demand for protein around the world means a great challenge to find sustainable alternative protein sources that have a high biological value [1]. Plants and green leaves have enormous potential for the production of protein concentrates.

Moringa oleifera is a foliated tree widely cultivated in tropical latitudes, which shows high adaptability to climatic conditions and dry soils [2]. It is a native species of South Asia, which grows at the foot of the Himalayas, from the northeast of Pakistan to the north of Bengal in India, presenting various nutritional, therapeutic or industrial benefits [3,4]. In particular, the leaves have been investigated as a valuable source of dietary protein and essential amino acids, and their use as an ingredient in livestock and human nutrition has been promoted [5].

In many parts of the world, including Africa, the use of *M. oleifera* as a nutritious food is on the rise. However, in developed countries, its use as a food ingredient is not well known. Its cultivation is not widespread in these regions either, despite its easy adaptation to hot climates (for example, the Mediterranean area). For that reason, this study was focused on the compositional and physicochemical analysis of different types of moringa leaves grown in Valencia. In addition, and in view of their possible incorporation into food matrices, the dry leaves were also characterized.

2. Materials and Methods

Three types of *Moringa oleifera* with different origins were used as raw material: Thailand (C1), Ghana (C2) and India (C3). They were grown in an experimental plot of the Universitat Politècnica de València, and their leaves were collected at the end of September 2020. The C1 and C2 types had cultural practices in the plants intended for leaf production, while the C3 was prepared to produce pods. Part of the leaves were dehydrated in a tray dryer with hot air at 50 °C for 8 h. Then, they were pulverized in a grinder (Thermomix, TM31, Vorwerk, Wupertal, Germany) for 3 min. The powder was sieved with a 0.5 mm sieve and stored in glass jars wrapped in aluminum foil and stored on dark shelves at room temperature for further experiments.

Water Content, Antioxidant Capacity, Optical Properties and Amino Acid Content

Water content was obtained by a gravimetric method (AOAC, 20.013, 2000) until a constant weight of the samples was reached after drying at 60 °C (VACIOTEM-T, J.P. Selecta Spain). An adaptation of the spectrophotometric DPPH method [6] was used to analyze the antioxidant capacity.

Color of moringa fresh leaves and powder was determined using a spectrocolorimeter (Konica Minolta, Inc., model CM-3600d, Tokyo, Japan). The results were expressed according to the CIE L*a*b* reference system with the D65 Standard Illuminant and 10° Standard Observer.

An EZ-Faast amino acid kit (Phenomenex, Torrance, CA, USA) was used to carry out the amino acid analyses [7].

3. Results and Discussion

The water content of fresh moringa leaves was 65 ± 3%, without significant differences between them. After drying (Figure 1A), the moringa of Thai origin (C1) registered the highest value of water content, with no significant differences from the other types of moringa (C2 and C3). In this sense, C1 requires a longer drying time despite having a vegetative level similar to that of C2, since both are used for the production of leaves, while moringa C3 is of the arboreal type for the production of pods.

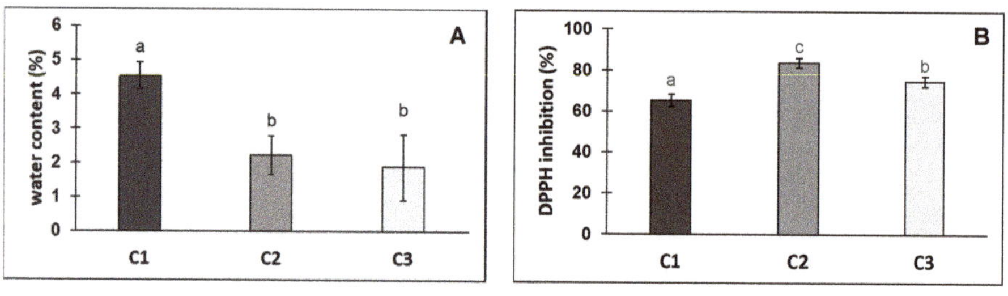

Figure 1. (**A**) water content and (**B**) antioxidant activity expressed as percentage of DPPH inhibition for different dehydrated moringa leaves (C1, C2 and C3). Equal letters indicate correlations coefficients statistically different ($p < 0.01$).

In all cases, DPPH inhibition (Figure 1B) was greater than 60%, highlighting moringa C2 (83 ± 2%), with similar antioxidant capacity reported in the Mexican [8] and fresh

Indonesian leaves. Therefore, the origin of plants would mainly influence the concentration of bioactive compounds and their antioxidant capacity.

3.1. Optical Properties

Figure 2 show a* and b* coordinates location in the chromatic plane of the fresh and dried leaves. As can be seen, regardless of the origin, the moringa powder presented greater purity of color than the fresh leaves because of the higher value in the coordinate b*. This behavior shows that the drying stage involves the loss of green tones, enhancing the yellows, due to the chlorophyll degradation and the formation of pheophytins [9]. In relation to the differences registered between the upper side and the underside of the leaves, as expected, the upper side presented higher values of a* and lower values of b*, giving evidence of coloration with a darker greenish intensity than that of the underside.

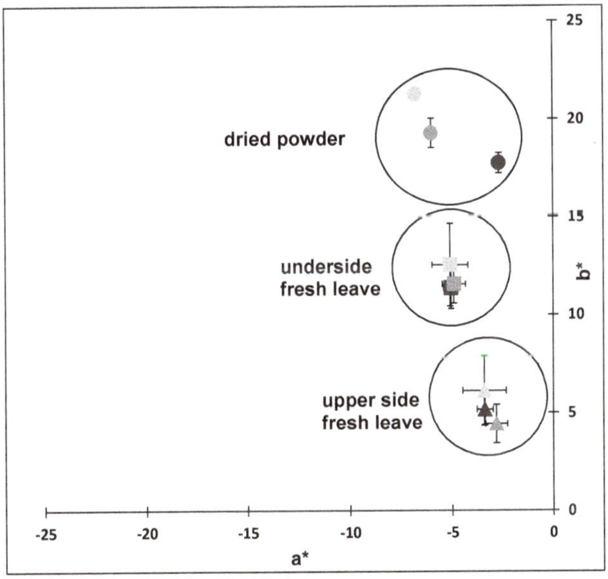

Figure 2. a* and b* coordinates location in the chromatic plane of the fresh and dried leaves. Dried powder: •C1, •C2, •C3; upper side fresh leaves: ▲C1, ▲C2, ▲C3; underside fresh leaves: ■C1, ■C2, ■C3.

3.2. Amino Acid Content

These moringa powders contained seven essential amino acids and were very rich in serine, glutamic acid and alanine without showing significant differences between the evaluated samples (Figure 3). The amount of glutamic acid content was similar to the reported value in dried moringa leaves grown in Greece [10] and South Africa [11]. However, the high serine content in these samples contrasted with concentrations found in other studies, where it was not one of the majority amino acids [10–12]. Furthermore, moringa intended for pod production (C3) showed the highest concentrations in some of the minor amino acids (threonine and asparagine). The high concentration of hydrophobic amino acids (alanine, leucine, isoleucine, phenylalanine and methionine) and acidic amino acid (glutamic acid) could be related to the high antioxidant power of moringa leaves [13].

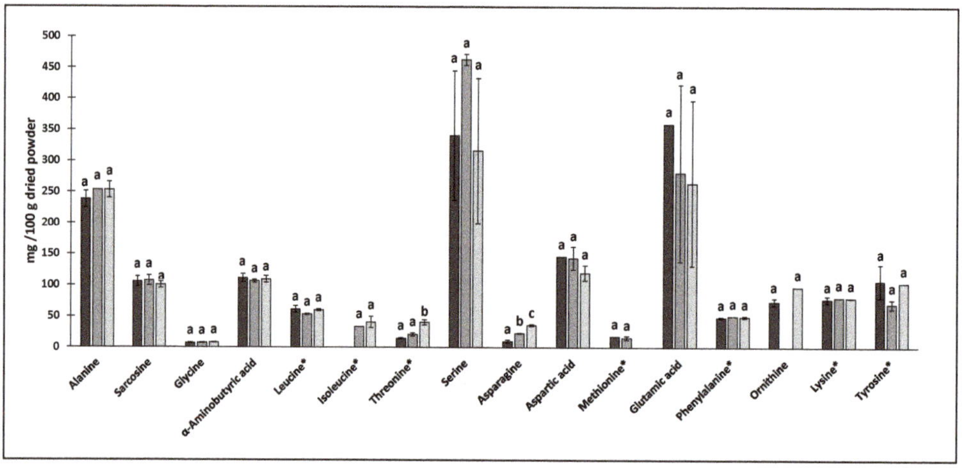

Figure 3. Amino acid content (* essential) of moringa powder: C1 (dark grey), C2 (intermediate grey) and C2 (light gray). Equal letters for the same amino acid indicate homogenous group in ANOVA ($p < 0.01$).

4. Conclusions

Fresh moringa leaves studied from seeds of different origins and with different cultural practices did not show significant differences in optical properties. Once dehydrated, the powders obtained did not show significant differences in the amino acid profile either, the most abundant being serine, glutamic acid and alanine. However, the plants destined for the production of leaves from Ghana (C2) had a higher antioxidant capacity. In addition, the moringa powder showed greater purity of color than the fresh leaves due to the loss of chlorophyll, although the greenish coloration persisted.

Institutional Review Board Statement: Not applicable.

Informed Consent Statement: Not applicable.

Acknowledgments: This work was supported by the grant Ia ValSe-Food-CYTED (119RT0567) and project Nuevos Cultivos Frente al Cambio climático: Moringa y Estevia (MORESVIA) (AG-COOP_A/2018/026) funded by Generalitat Valenciana.

References

1. McCarty, M.F.; Barroso-Aranda, J.; Contreras, F. The low-methionine content of vegan diets may make methionine restriction feasible as a life extension strategy. *Med. Hypotheses* **2009**, *72*, 125–128. [CrossRef] [PubMed]
2. Okuda, T.; Baes, A.U.; Nishijima, W.; Okada, M. Isolation and characterization of coagulant extracted from Moringa oleifera seed by salt solution. *Water Res.* **2001**, *35*, 405–410. [CrossRef]
3. Aderinola, T.A.; Alashi, A.M.; Nwachukwu, I.D.; Fagbemi, T.N.; Enujiugha, V.N.; Aluko, R.E. In vitro digestibility, structural and functional properties of Moringa oleifera seed proteins. *Food Hydrocoll.* **2020**, *101*, 105574. [CrossRef]
4. Nouman, W.; Anwar, F.; Gull, T.; Newton, A.; Rosa, E.; Domínguez-Perles, R. Profiling of polyphenolics, nutrients and antioxidant potential of germplasm's leaves from seven cultivars of Moringa oleifera Lam. *Ind. Crops Prod.* **2016**, *83*, 166–176. [CrossRef]
5. Amaglo, N.K.; Bennett, R.N.; Lo Curto, R.B.; Rosa, E.A.S.; Lo Turco, V.; Giuffrida, A.; Lo Curto, A.; Crea, F.; Timpo, G.M. Profiling selected phytochemicals and nutrients in different tissues of the multipurpose tree Moringa oleifera L., grown in Ghana. *Food Chem.* **2010**, *122*, 1047–1054. [CrossRef]
6. Turkmen, N.; Sari, F.; Velioglu, Y.S. The effect of cooking methods on total phenolics and antioxidant activity of selected green vegetables. *Food Chem.* **2005**, *93*, 713–718. [CrossRef]
7. Periche, A.; Koutsidis, G.; Escriche, I. Composition of antioxidants and amino acids in stevia leaf infusions. *Plant Foods Hum. Nutr.* **2014**, *69*, 1–7. [CrossRef] [PubMed]
8. Cuellar-Nuñez, M.L.; Luzardo-Ocampo, I.; Campos-Vega, R.; Gallegos-Corona, M.A.; González de Mejía, E.; Loarca-Piña, G. Physicochemical and nutraceutical properties of moringa (*Moringa oleifera*) leaves and their effects in an in vivo AOM/DSS-induced colorectal carcinogenesis model. *Food Res. Int.* **2018**, *105*, 159–168. [CrossRef] [PubMed]

9. Sledz, M.; Witrowa-Rajchert, D. Influence of microwave-convective drying of chlorophyll content and colour of herbs. *Acta Agrophysica* **2012**, *19*, 865–876.
10. Lalas, S.; Athanasiadis, V.; Karageorgou, I.; Batra, G.; Nanos, G.D.; Makris, D.P. Nutritional characterization of leaves and herbal tea of moringa oleifera cultivated in Greece. *J. Herbs Spices Med. Plants* **2017**, *23*, 320–333. [CrossRef]
11. Moyo, B.; Masika, P.J.; Hugo, A.; Muchenje, V. Nutritional characterization of Moringa (*Moringa oleifera* Lam.) leaves. *Afr. J. Biotechnol.* **2011**, *10*, 12925–12933. [CrossRef]
12. Castillo-López, R.I.; León-Félix, J.; Angulo-Escalante, M.Á.; Gutiérrez-Dorado, R.; Muy-Rangel, M.D.; Heredia, J.B. Nutritional and phenolic characterization of moringa Oleifera leaves grown in Sinaloa, México. *Pakistan J. Bot.* **2017**, *49*, 161–168.
13. Natsir, H.; Wahab, A.W.; Budi, P.; Dali, S.; Arif, A.R. Amino acid and mineral composition of moringa oleivera leaves extract and its bioactivity as antioxidant. *J. Phys. Conf. Ser.* **2019**, *1317*, 012030. [CrossRef]

biology and life sciences forum

MDPI

Proceeding Paper

Gluten-Free Breadmaking with Extruded Whole-Grain Andean Maize Flours †

Rita M. Miranda, Argentina A. Amaya, Manuel O. Lobo and Norma C. Sammán *

Centro de Investigación Interdisciplinario en Tecnología y Desarrollo Social del NOA (CIITED), Consejo Nacional de Investigaciones Científicas y Técnicas (CONICET) Facultad de Ingenierí, Universidad Nacional de Jujuy, Ítalo Palanca 10, San Salvador de Jujuy 4600, Argentina; ritamiranda89@gmail.com (R.M.M.); argentinaamaya12@gmail.com (A.A.A.); mlobo958@gmail.com (M.O.L.)

* Correspondence: normasamman@gmail.com

† Presented at the III Conference la ValSe-Food and VI Symposium Chia-Link Network, Online, 15–17 November 2021.

Abstract: Andean maize can be safely used in gluten-free bread formulation. Extrusion is a technology capable of promoting changes in the techno-functional properties of gluten-free flours, modifying their breadmaking properties. The objective of this study was to evaluate the effect of extrusion on the physical and physicochemical properties of Andean maize whole-grain flours (*bolita* race) and to determine the relationship between the changes to the textural properties of gluten-free dough and bread with the addition of extruded flours. The Andean maize whole-grain flours were extruded in a single-screw extruder. The moisture, temperature and screw speed were varied through an incomplete orthogonal design. The expansion degree of extruded products, the total soluble carbohydrates, and the gelatinization degree of the flours varied mainly with moisture and temperature extrusion. Flours with high, medium, and low degrees of gelatinization treatments were added at 20 % to native flours to make gluten-free dough and bread. The dough made with the addition of extruded flours increased their firmness and adhesiveness in relation to the control made with native flour alone. Bread made with extruded flours generally increased their hardness, gumminess, chewiness, and cohesiveness. Springiness only increased under conditions of high and low degrees of gelatinization. The dough made with extruded flour at the extruded condition of 100 °C-25%H-120 rpm, with the lowest degree of gelatinization, were the least firm and adhesive, which could lead to better dough machinability. Additionally, the bread made with this flour presented high cohesiveness and springiness.

Keywords: Andean maize; bread; extrusion; gluten-free; whole-grain flours

Citation: Miranda, R.M.; Amaya, A.A.; Lobo, M.O.; Sammán, N.C. Gluten-Free Breadmaking with Extruded Whole-Grain Andean Maize Flours. *Biol. Life Sci. Forum* **2021**, *8*, 5. https://doi.org/10.3390/blsf2021008005

Academic Editors: Loreto Muñoz and Claudia M. Haros

Published: 16 December 2021

1. Introduction

In the Puna and Quebrada of Humahuaca, there are several varieties of Andean maize which possess different nutritional and physicochemical properties. Andean maize can be used in the production of gluten-free foods because of the absence of prolamines, a gluten-forming protein fraction that affects celiac patients [1]. Generally, gluten-free bread has defects such as low volume and poor texture because of the gluten absence; gluten acts as a structural support that allows the dough to expand during proofing [2]. Additionally, gluten-free bread is usually formulated with refined flours or starches with low nutritional value, to which technological enhancers such as hydrocolloids are added [3]. Therefore, gluten substitution continues to be a challenge for food technologists.

Whole-grain flours compared to refined ones have more nutrients and dietary fiber. The type of fiber present in flours can affect the stability of the dough and make it difficult to obtain gluten-free bread of good technological quality [4]. Extrusion cooking has been used to modify the techno-functional properties of starchy and high-fiber raw materials. This technology has been applied to gluten-free flours to obtain technologically improved bread with acceptable results [5]. However, few have studied the effects of extrusion on the breadmaking properties of Andean whole-grain flours.

The objective of this study was to evaluate the effect of extrusion on the physical and physicochemical properties of whole-grain flours of Andean maize (*bolita* race) and to determine the relationship of these with the changes in textural properties of gluten-free dough and bread due to the addition of extruded flours.

2. Materials and Methods

2.1. Materials

Andean maize (*bolita* race) grown in the Ocumazo-Humahuaca province of Jujuy was used. The grains were milled in a hammer mill to obtain whole-grain flour with grain size <710 μm. The whole-grain flour was extruded in a Brabender single-screw extruder with a compression ratio of 3:1, 3 mm nozzle, at temperatures of 100, 120, and 140 °C, moistures of 15, 20, and 25%, and screw speeds of 80, 100, and 120 rpm, using an incomplete orthogonal design.

2.2. Extruded Material Characterization

The expansion degree (ED) of the extruded products was determined as the ratio of the diameter of the expanded product and the diameter of the extruder nozzle [6]. Ten replicates of diameters of the expanded products were determined.

The total soluble carbohydrates (TSC) of the extruded flours suspension supernatants were measured using the phenol-sulfuric method according to Taylor (1995) with modifications in the preparation of samples [7]. A calibration curve was established using pure xylose solutions as standards, processed by the same procedures. The measurement was carried out in triplicate.

The gelatinization degree (GD) of the extruded flours was determined according to the method of Baks (2007) with modifications in the preparation of samples [8]. The measurement was carried out in triplicate.

2.3. Bread Preparation

The lactal bread was made with whole-grain native maize flour substituted with 20% extruded flour; initially, the flours were mixed for 1 min to achieve homogenization of the samples. For every 100 g of substituted flours, 110 mL of spout water (30 °C), 1 g of previously activated commercial dry yeast, 1 g of salt (NaCl), 2 g of sugar, 3 g of powdered milk, and 6 mL of oil were added and mixed at low speed for 5 min. The dough obtained was put in molds and placed in a fermentation chamber 50 min at 30 °C and 80–90% relative humidity. The fermented dough was baked at 150 °C for 50 min, and textural properties were immediately determined. The bread was evaluated 24 h after baking.

2.4. Textural Properties of Gluten-Free Dough and Bread

The gluten-free fermented dough and bread texture profile analysis (TPA) was performed using a texture analyzer (TAXT plus, Stable Micro System, Godalming, UK) equipped with a 5 kg load cell. A Teflon cylindrical probe with a P/0.5 (12.7 mm) was used for dough, and 25% deformation and a 20.0 s waiting time were used. The test speed was set to 5.0 mm/s. An aluminum cylindrical probe with a P/35 (35.0 mm) was used for bread; samples from the center of the bread (thickness of 10 mm) were compressed to 50% of their original height. The test speed was 1 mm/s and the waiting time was 5.0 s. The measurements were made in quadruplicate.

2.5. Statistical Analysis

Data were analyzed using INFOSTAT software. The significant difference between the means was evaluated by Tukey's test ($p < 0.05$) using analysis of variance (ANOVA). Data on the textural properties of the gluten-free dough and bread formulated with the addition of extruded flours (preselected conditions) were summarized in a principal component analysis.

3. Results and Discussion

3.1. Properties Physical and Physicochemical of the Extruded Material

Figure 1A shows the ED of the maize (*bolita* race) extrudates; it varied between 1.68 ± 0.24 and 2.74 ± 0.21, with the highest value at 100 °C-15% H-80 rpm. ED showed a tendency to decrease with the increase in extrusion moisture at different temperatures. This variation was only significantly different between extruded samples at the same moisture between 15 and 20% and 100 and 120 °C; the most important effect was the moisture. A similar trend was found by Byars (2015) in mixtures of degermied corn flour with chia [6].

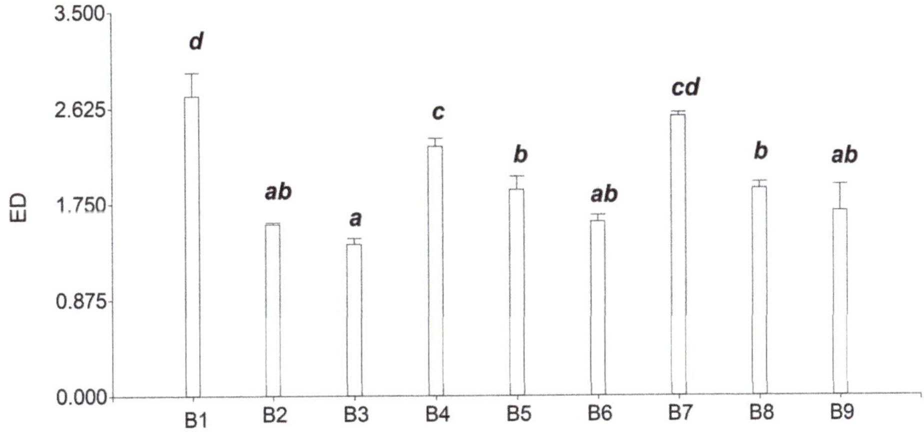

(A)

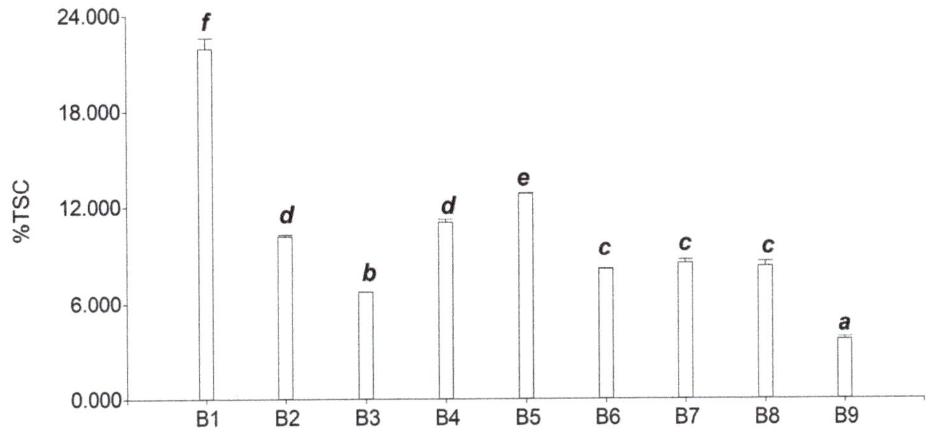

(B)

Figure 1. *Cont.*

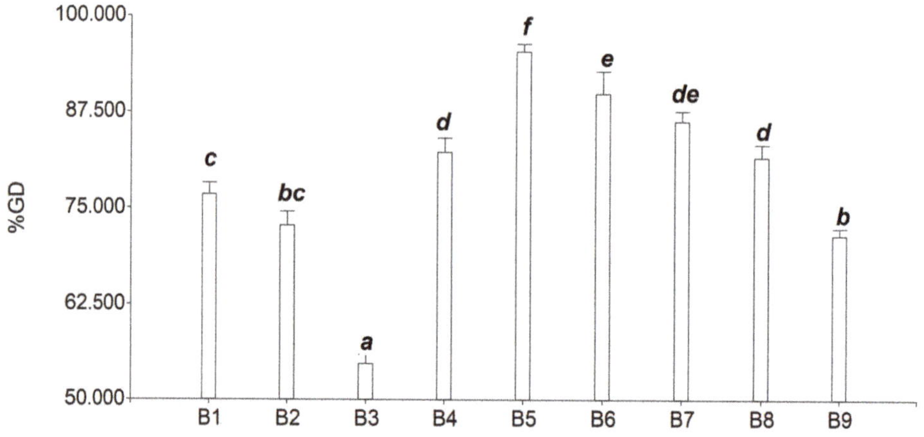

(C)

Figure 1. (**A**) Expansion degree of extruded products. (**B**) Total soluble carbohydrates of the supernatants and (**C**) gelatinization degree of the extruded flours of Andean maize (*Bolita* race). B1:100 °C-15%H-80 rpm; B2:100 °C-20%H-100 rpm; B3:100 °C-25%H-120 rpm; B4: 120 °C-15%H-100 rpm; B5:120 °C-20%H-120 rpm; B6:120 °C-25%H-80 rpm; B7:140 °C-15%H-120 rpm; B8:140 °C-20%H-80 rpm; B9:140 °C-25%H-100 rpm.

Figure 1B shows the variation of the content of the TSC present in the supernatant of suspensions of extruded flours in water (3.68 ± 0.18 to 21.95 ± 0.68%). The extruded flour at 100 °C-15% H-80 rpm presented the highest content of TSC. A tendency to decrease the content of TSC was observed with increasing extrusion moisture, which was more pronounced at 100 °C. The difference was significant between extruded samples at the same moisture level at the different temperatures, which suggests a greater influence of the extrusion temperature on this parameter. The TSC content was not affected by screw speed.

The content of TSC in the supernatant of suspensions of flours extruded in water was determined due to the increase in the water solubility index of these samples with respect to the native flour (data not shown). The flour extruded at low extrusion temperatures (100 °C and 120 °C) had a high DE and a higher content of TSC in the supernatants of the suspensions. This indicates the degradation of complex carbohydrates during extrusion. These results agree with those of Sarifudin (2014), who indicated that there is greater degradation at low extrusion moisture due to shear [9].

The GD of the extruded flours is shown in Figure 1C. This parameter increased at equal moisture levels at temperatures of 100 and 120 °C. The highest GD was observed at 120 and 140 °C; the greatest was at 120 °C-20% H-120 rpm. No trend was observed regarding the variation of this parameter with screw speed.

To carry out the breadmaking tests, the extruded samples that were significantly different in terms of the parameters studied were selected: B1 (100 °C-15%H-80 rpm) with an intermediate GD, high TSC content, and high ED; B3 (100 °C-25%H-120 rpm) with a low GD, low TSC content, and low ED; B5 (120 °C-20%H-120 rpm) with a high GD, intermediate TSC content, and intermediate ED.

3.2. Textural Characteristics of the Doughs and Bread Gluten Free

Figure 2 shows the principal component analysis of the physical–physicochemical properties of the extruded flours B1, B3, and B5, and the textural properties determined in the dough and breads added with these flours. A positive correlation was observed between the adhesiveness of the dough, the ED, and the content of TSC. The hardness, springiness, gumminess, and chewiness showed a negative correlation with the ED and the content of TSC. The cohesiveness showed a negative correlation with the GD. In general, the GD did not show a correlation with the textural properties.

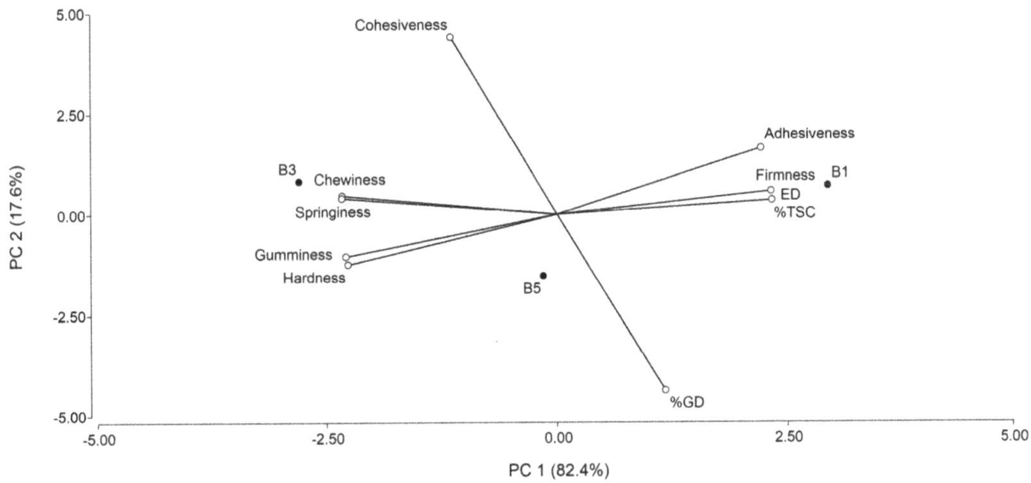

Figure 2. Principal component analysis of textural properties of the gluten-free doughs and breads made from native flours added at 20 % of the B1:100 °C-15%H-80 rpm, B3:100 °C-25%H-120 rpm, and B5:120 °C-20%H-120 rpm.

The dough increased its firmness between 67.7 and 91.9% compared to the native flour control (9.93 g). The firmness was higher in the dough made with extruded flours with contents high in TSC and intermediate ED (B1 and B5); the dough made with flour with a TSC and a low ED (B3) presented the lowest firmness of the three preselected flours. The adhesiveness increased between 129.5 and 268.9% in relation to the control (3.73 gs); the dough made with B3 showed the lowest value. The dough made with B3 presented textural characteristics that could facilitate its handling during the breadmaking process.

Breads made with the addition of extruded flours increased their hardness (107.4–194.7%), cohesiveness (36.6–43.3%), gumminess (78.3–163.6%), and chewiness (54–186.9%) with respect to the native flour control. Condition B3 presented the highest increases in the parameters mentioned above. Springiness increased by 10.3% in this condition alone compared to the control. Greater cohesion and elasticity could positively contribute to the acceptability of bread by consumers.

4. Conclusions

Extrusion modified the physical and physicochemical properties of whole-grain Andean maize (*bolita* race) flours, with a greater effect of moisture and temperature of processing. In general, the textural properties of the formulated dough and bread showed a correlation with the expansion degree and the content of total soluble carbohydrates. The selection of low temperature and high moisture extrusion conditions, such as 100 °C and 25%H, would positively contribute to the industrial handling of the dough and to the acceptability of the baked product.

Institutional Review Board Statement: Not applicable.

Informed Consent Statement: Not applicable.

Acknowledgments: This work was supported by grant Ia ValSe-Food-CYTED (Ref. 119RT0567), CONICET and SECTER, National University of Jujuy (Argentina).

Conflicts of Interest: The authors declare no conflict of interest.

References

1. Giménez, M.A.; Gámbaro, A.; Miraballes, M.; Roascio, A.; Amarillo, M.; Sammán, N.; Lobo, M. Sensory evaluation and acceptability of gluten-free Andean maize spaghetti. *J. Sci. Food Agric.* **2015**, *95*, 186–192. [CrossRef] [PubMed]
2. Sciarini, L.S.; Steffolani, M.E.; León, A.E. El rol del gluten en la panificación y el desafío de prescindir de su aporte en la elaboración de pan. *AgriScientia* **2016**, *33*, 61–74. [CrossRef]
3. Mir, S.A.; Shah, M.A.; Naik, H.R.; Zargar, I.A. Influence of hydrocolloids on dough handling and technological properties of gluten-free breads. *Trends Food Sci. Technol.* **2016**, *51*, 49–57. [CrossRef]
4. Tsatsaragkou, K.; Protonotariou, S.; Mandala, I. Structural role of fibre addition to increase knowledge of non-gluten bread. *J. Cereal Sci.* **2016**, *67*, 58–67. [CrossRef]
5. Gómez, M.; Martínez, M.M. Changing flour functionality through physical treatments for the production of gluten-free baking goods. *J. Cereal Sci.* **2016**, *67*, 68–74. [CrossRef]
6. Byars, J.A.; Singh, M. Properties of extruded chia-maize meal puffs. *LWT* **2015**, *62*, 506–510. [CrossRef]
7. Al Loman, A.; Islam, S.M.; Li, Q.; Ju, L.K. Enzyme recycle and fed-batch addition for high-productivity soybean flour processing to produce enriched soy protein and concentrated hydrolysate of fermentable sugars. *Bioresour. Technol.* **2017**, *241*, 252–261. [CrossRef] [PubMed]
8. Baks, T.; Ngene, I.S.; van Soest, J.J.G.; Janssen, A.E.M.; Boom, R.M. Comparison of methods to determine the degree of gelatinisation for both high and low starch concentrations. *Carbohydr. Polym.* **2007**, *67*, 481–490. [CrossRef]
9. Sarifudin, A.; Assiry, A.M. Some physicochemical properties of dextrin produced by extrusion process. *J. Saudi Soc. Agric. Sci.* **2014**, *13*, 100–106. [CrossRef]

biology and life sciences forum

MDPI

Proceeding Paper

Development of a Latin American Native Food Composition Database †

Natalia Bassett [1,2], Constanza Rossi [1,2], Argentina Amaya [1], Francisco Ríos [1] and Norma Sammán [1,2,*]

1 Centro Interdisciplinario de Investigaciones, Tecnologías y Desarrollo Social para el NOA (CIITED), Consejo Nacional de Investigaciones Científicas y Técnicas (CONICET), Universidad Nacional de Jujuy, San Salvador de Jujuy 4600, Argentina; natybassett@gmail.com (N.B.); cotyrossi@hotmail.com (C.R.); argentinaamaya12@gmail.com (A.A.); fran_stal@hotmail.com (F.R.)

2 Instituto Superior de Investigaciones Biológicas (INSIBIO), CONICET, Universidad Nacional de Tucumán, San Miguel de Tucumán 4000, Argentina

* Correspondence: normasamman@gmail.com

† Presented at the III Conference la ValSe-Food and VI Symposium Chia-Link Network, online, 15–17 November 2021.

Abstract: Food composition data have a fundamental function in studies on nutrition, health, and agriculture, among others. Many factors affect the nutrient content of food, and for this reason, it is essential to have updated and reliable data on the composition of the main foods consumed. The objective of this work was to develop a food composition database (FCDB) that compiles the composition of native foods of Latin America, mainly grains/seeds, tubers, and derivatives. An interdisciplinary work group of compilers was formed. A search of various sources was carried out (scientific publications, laboratory and technical reports, and theses), and a total of 78 publications were collected. For compilation, a form composed of eight worksheets was prepared. The initial sheet contains general data and food identification; the remaining ones contain information on the proximal composition, amino acids, fatty acids, vitamins, and minerals. Each section has an evaluation of data quality, which determines whether it will be included in the FCDB or not. After an exhaustive analysis based on compliance with the minimum requirements previously established, 58 publications and laboratory reports were selected. The main reason for rejection was the lack of moisture information (60%), followed by low data quality (30%). Information is available on the composition of at least 26 grains and derived products (i.e., quinoa, amaranth, and kañiwa) and five tubers and roots (Andean potatoes and ocas), which are currently being uploaded to the website (http://insibio.org.ar, accessed on 16 December 2021) for user availability. This database will provide information on the composition of regional foods generated and compiled using international standards.

Keywords: database; food; Latin American; native; grains

Citation: Bassett, N.; Rossi, C.; Amaya, A.; Ríos, F.; Sammán, N. Development of a Latin American Native Food Composition Database. *Biol. Life Sci. Forum* **2021**, *8*, 6. https://doi.org/10.3390/blsf2021008006

Academic Editors: Loreto Muñoz and Claudia M. Haros

Published: 16 December 2021

1. Introduction

The Food and Agriculture Organization of the United Nations (FAO) actively promotes the conservation and sustainable use of biodiversity for nutrition and agriculture as a means of increasing dietary diversity. However, at present, there is a strong tendency to reduce the base of global food security to a few species. This decline reduces the ability of farmers and ecosystems to adapt to changes and opportunities that arise. From a research point of view, solving this situation requires a greater range of species. This involves revaluing and including those that are used by populations and communities. In the case of the Andean region, underutilized species are adapted to the agroecological conditions of the region, are used by local farmers, and thereby contribute to sustainable production and the stability of the ecosystem. However, there is limited knowledge of the nutritional values of these species and a lack of strategies for their inclusion in food and nutrition programs. Food composition data have a fundamental function in studies on nutrition, health, and agriculture, among others. Biodiversity is a concept that summarizes the different kinds of organisms in the biosphere, namely, plants and animals, as well as genes that have the organisms and their habitat. Many factors affect the nutrient content of food, and for this reason, it is essential to have updated and reliable data on the composition of the main foods consumed. This can be achieved through the preparation and use of a food composition database (BDCA) and/or food composition tables (TCA) regionally and nationally.

The Andes, with about 8000 km of travel on the western side of South America from Venezuela to Patagonia, Argentina, are important geological heterogeneity ecological life zones and host biodiversity. The most prominent influence of the Andes is the plant biodiversity and cultural aspects. Some examples of crops typical of the native Andean foods are Andean potatoes, oca, yacón, amaranth, cañihua, tarwi, and quinoa, among others. Although these crops are native to different areas of Latin America, their cultivation and consumption have spread to other areas of the world [1].

The process of preparing a table must follow standardized procedures, which guarantee a quality final product. That is why the Food and Agriculture Organization of the United Nations (FAO) has made available a series of documents aimed at standardizing production processes and a compilation of chemical composition data in order to maximize the quality of the tables obtained [2].

Within this framework, the objective of this work is to report the process of preparing an updated and methodologically adequate Table of Food Composition of Native Foods of Latin American, which responds, among others, to the needs of food programs and to the formulation of new healthy foods [3,4]. It is expected that this BDCA contributes to the knowledge of forgotten or underutilized crops and, hence, contributes to their recovery and revaluation; provides nutrition researchers with estimates of the contribution of biodiversity to nutrition; allows agricultural researchers to select those crops or breeds that have a high-quality nutritional profile for agricultural research, dissemination, and larger scale production.

2. Materials and Methods

2.1. Description

The present work was based on the systematic compilation of food composition data, both analytical and published or from laboratory reports, which was followed by a methodological unification of the information compiled to develop a homogeneous database. For this reason, this research is descriptive, cross-sectional, and retrospective.

To carry out the objective, a working group of technicians and professionals was formed. It was important to train compilers in the general principles of compilation, organization of databases, and preparation of food composition tables, as well as in knowledge of food sampling plans and analytical methods for data generation and quality verification, thus forming the criteria for the selection of the material that is to be incorporated into the BDCA.

2.2. Information Gathering

The database document repository is a collection of original online documents containing analytical data, either from the scientific literature (e.g., searched for in PubMed, ScienceDirect, or Scopus) or laboratory reports. All available information was taken as the universal or target population, including published works, graduate and postgraduate theses, laboratory reports, and papers presented and published at congresses and scientific meetings.

2.3. Criteria for Inclusion of the Information in the BDCA

The inclusion criteria of the search pointed to works that contained the following information:

- General information: name of the food; variety; a detailed description of the food; analyzed part; scientific name; trade name (for industrialized products); the number of samples analyzed; origin of the samples (geographic, local of acquisition); analytical methods used; bibliographic reference of the analytical method; origin of the information (laboratory that performed the analysis).
- Minimum food information: description of the handling of samples; humidity; percentage information of the nutrients analyzed; some index of variability as range or standard deviation; analytical quality control; date of production of the food analysis.

2.4. Food Groups

Three food groups were defined (A—cereals and derivatives; B—vegetables and legumes; C—seeds and derivatives) out of the 16 groups of LATINFOODS. Each group was assigned an identification letter.

The food description was formulated using the updated 2017 LanguaL System. LanguaL is a food description thesaurus that provides a standardized language using faceted classification. The descriptors include additional information (scientific names, references, synonyms, processes). It also includes the Codex Alimentarius classification [5–7].

2.5. Compilation Forms

The designed compilation form is composed of eight worksheets. The initial sheets contain general data and food identification; the remainder contain information on proximal composition, amino acids, fatty acids, water, and fat-soluble vitamins and minerals. Each food is classified according to the system adopted by LATINFOODS; the group is registered only with the letter to which it belongs, and then a three-digit numerical code is given to each food entry, which indicates the order within the group. The unit of measurement is g/100 g food on a wet basis (as consumed). For other nutrients, the units are those agreed in INFOODS and are explicitly indicated. In all cases, the humidity data must be included. Each section has an evaluation of data quality, which determines whether it will be included in the FCDB or not (Figure 1).

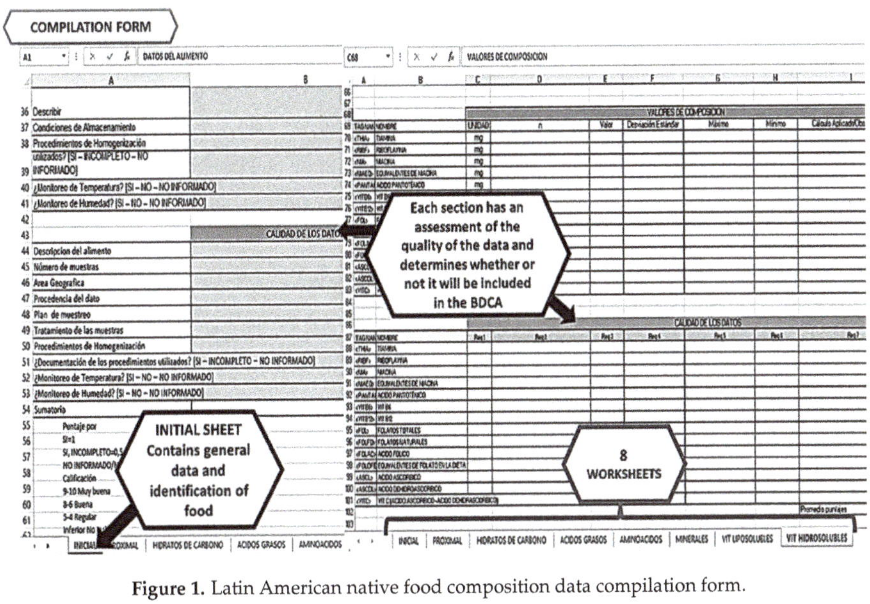

Figure 1. Latin American native food composition data compilation form.

2.6. User Manual Preparation

The analysis and the preparation of the user manual was carried out with the guides and steps necessary to complete the forms for the collection of native food composition data. In addition, the working group was trained on how to use them. Criteria were agreed on the type of data and description of the foods to be included in the compilation sheets.

3. Results

3.1. Number of Works Collected and Their Origins

After an exhaustive analysis based on compliance with the minimum requirements previously established, 78 publications and laboratory reports were selected. After the first step of the analysis of the compliance with minimum inclusion criteria and the evaluation of data quality, 43 publications were incorporated into the database (Figure 2). Some published works include more than one record for the analysis of the composition of varieties of grains and seeds and also of simple derivatives, such as flours (for this reason, the database currently has approximately 110 records, although it is updated every three months). In the case of Latin American native foods, it is very important to determine the variability within the same food because of the great biodiversity. Among the works selected to be part of the database, the main countries in Latin America to which the publications belong were Argentina, Mexico, Peru, and Ecuador. Outside Latin American territory, publications from Canada, Nigeria, and Portugal, among others, were selected. Regarding the period in which the works were published, these data are presented in Figure 3.

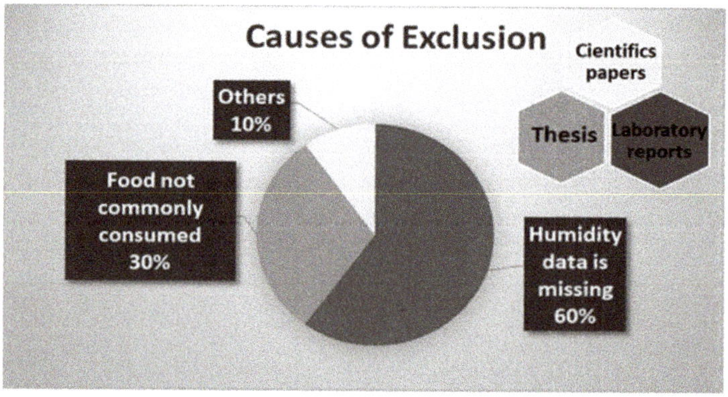

Figure 2. Causes of exclusion and distribution of data included in the BDCA.

3.2. Foods with more Frequency of Appearance

The database contains information on proximal composition only of vegetable origin, mostly raw, although some information on minimally processed products (flours) is included. Information is available on the composition of at least 26 grains and derived products (i.e., quinoa, amaranth, and kañiwa) and five tubers and roots (Andean potatoes and ocas). Moreover, also included are 11 legumes (chia and beans); 1 Sacha inchi; and 5 yellow, white, and purple maize, among others (Figure 4).

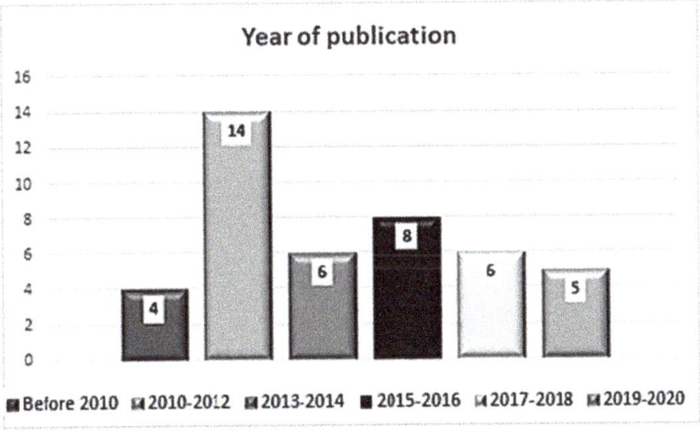

Figure 3. Year distribution of published composition data.

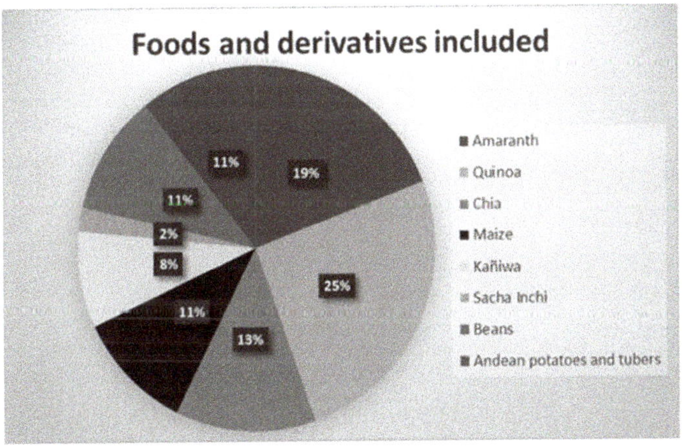

Figure 4. Foods and derivates included.

3.3. Causes of Rejection of Information and Data Quality

Of the total (N = 78) pieces of information collected, 43 (55.1%) were selected; the rest were rejected for different reasons mainly due to the low quality of the data. Table 1 and Figure 2 show the distribution of the data included in the BDCA and the causes of exclusion, respectively. The main reason for rejection was the lack of information of humidity data (60%), followed by food not commonly consumed (30%) and others (10%). Regarding the quality of the data of the publications included in the database, Figure 5 shows to which category the analyzed works belong.

Table 1. Amount of data accepted or rejected in the BDCA.

Accepted Data	Not Accepted Data	Rejected Data	Total
N = 43	N = 15	N = 20	N = 78
55.1%	19.2%	25.6%	100%

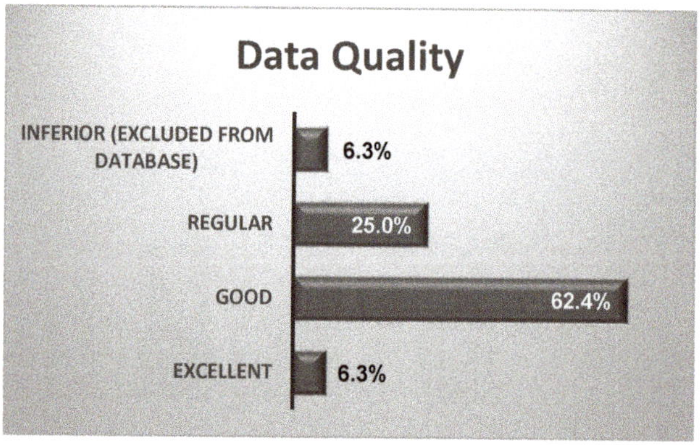

Figure 5. Data quality.

4. Discussion

In Latin America, most of the countries have developed TCAs, and although many of them are not available on the web, this information should be prioritized. In general, the amount of foods and micronutrients analyzed is low, so it is generally necessary to resort to other sources of information, such as TCAs and BDCAs, developed outside the region and that are more robust, such as those of the US, Denmark, and Canada. However, most of these BDCAs do not contain information on Andean native products, therefore indicating the importance of having a database with native foods of the region.

The working group held periodic meetings in which all the information was collected and distributed to achieve agreements, uniformity of criteria, and management of the compilation forms and how to complete them.

The information on the composition of the native foods entered into the BDCA refers mainly to macronutrients; there are few works with water-soluble and fat-soluble vitamin data. Some races of maize and varieties of chia were entered as average data taking into account the content of the main nutrient, for example, protein or lipid content; all entries are clearly explained in the section *Observations* of the worksheets. In other cases, the differences in the content of some macronutrients, such as proteins, were so significant that the data were entered into the BCDA as an individual record for each variety. In general, there is also very little information on the content of fatty acids and amino acids in these native Latin American foods. Quinoa and amaranth seeds are the food items that present the most complete data.

Regarding the number of publications found in the systematic search, among the native grains with the highest number of publications were quinoa and amaranth. To a lesser extent, data were found on kañiwa, chia, and legumes. Finally, Sacha inchi was the food with the fewest records. It could also be observed that the publications referring to these crops were made mainly between the years 2010 and 2012 and 2015 and 2016. Currently, there is a decrease in scientific publications and reports on the composition of these autochthonous foods.

This database is available to users at the following link: http://insibio.org.ar/.

5. Conclusions

A database is a tool that provides an affordable way to manage information on the composition of native foods generated and compiled using international standards, which are currently displayed in different types of scientific work.

This BDCA will contribute to bringing new insights into the research field of generally underutilized genetic resources of biodiversity in Latin America, which could be used as non-conventional food.

A future goal is to continue the process of compilation, improving the harmonization and standardization of composition data to allow their comparison and interchange. Composition data for recipes and meals will also be compiled.

Institutional Review Board Statement: Not applicable, since this study does not involve humans.

Informed Consent Statement: Not applicable.

Data Availability Statement: The data presented in this study are available in figures and tables within this article.

Acknowledgments: This work was supported by grant Ia ValSe-Food-CYTED (119RT0567) and CIITED-CONICET.

References

1. FAO. *The State of the World's Biodiversity for Food and Agriculture*; Bélanger, J., Pilling, D., Eds.; FAO Commission on Genetic Resources for Food and Agriculture Assessments: Rome, Italy, 2019; 572p, Available online: http://www.fao.org/3/CA3129EN/CA3129EN.pdf (accessed on 5 October 2021).
2. OMS/FAO; Organización de las Naciones Unidas para la Alimentación y la Agricultura. International Network of Food Data Systems (INFOODS). Available online: http://www.fao.org/infoods/ (accessed on 5 October 2021).
3. Sammán, N.; Rossi, M.C.; Ríos, F.; Romero, I.; Levy, L.; Tenisi, M.; Barrios, R. Desarrollo de una tabla nacional de composición de alimentos para diseñar políticas y programas de seguridad alimentaria. *Rev. Argent. Salud Pública* **2020**, *12*, 25.
4. Sammán, N.; Masson, L.; de Pablo, S.; Ovelar, E. Food composition activities in Argentina, Chile and Paraguay. *J. Food Compos. Anal.* **2011**, *24*, 716–719. [CrossRef]
5. Charrondiere, U.R.; Burlingame, B. Report on the FAO/INFOODS Compilation Tool: A simple system to manage food composition data. *J. Food Compos. Anal.* **2011**, *24*, 711–715. [CrossRef]
6. Ireland, J.D.; Møller, A. LanguaL food description: A learning process. *Eur. J. Clin. Nutr.* **2010**, *64*, S44–S48. [CrossRef] [PubMed]
7. LanguaL. The International Framework for Food Description. 2017. Available online: http://www.langual.org (accessed on 5 October 2021).

biology and life sciences forum

MDPI

Proceeding Paper

Chia Seed Oil Intake: Is It Beneficial for Preventing Cardiovascular Risk Factors? †

Susana Jerez [1,2,*], Analía Medina [2], Gabriela Alarcón [1], Liliana Sierra [2] and Mirta Medina [2]

1 Instituto Superior de Investigaciones Biológicas (INSIBIO-CONICET-UNT), San Miguel de Tucuman 4000, Argentina; gabydel.jesus@gmail.com
2 Facultad de Ciencias Naturales e IML-UNT, San Miguel de Tucuman 4000, Argentina; analiamed@csnat.unt.edu.ar (A.M.); lilianasierra@csnat.unt.edu.ar (L.S.); medmirta@csnat.unt.edu.ar (M.M.)
* Correspondence: sjerez@herrera.unt.edu.ar
† Presented at the III Conference la ValSe-Food and VI Symposium Chia-Link Network, Online, 15–17 November 2021.

Abstract: Cold-pressed chia seed oils (ChO) are known for their health-promoting characteristics due to their high content of omega-3 α-linolenic acid (ω-3 ALA). We investigated the influence of ChO supplementation as a functional food on animal models of the cardiovascular risk factors hypercholesterolemia and metabolic syndrome (MS). Dietary intervention with ChO (equivalent to 4.8 g ALA per day) was found to improve vascular dysfunction and mitigate the rise in plasma triglyceride (TG) levels under hypercholesterolemic conditions. However, impaired glucose tolerance was found in control ChO-treated animals. In order to verify whether the effects of chia seed are the same as that of ChO, we replaced ChO with an equivalent amount of seed. Glucose intolerance was found once again. For this reason, we carried out a new study in which ChO intake was reduced to 3 g ALA per day, and no alterations were observed in such conditions. Thus, dietary intervention with ChO equivalent to 3 g ALA intake per day was chosen to analyze the effects on the alterations that characterize high-fat diet-induced MS. ChO supplementation lowered the ω-6/ω-3 ratio, TG, blood pressure and improved endothelial function. Nevertheless, ChO worsened the high-fat diet's deleterious effects on visceral abdominal fat, fasting glucose and glucose tolerance. Our results support the view that dietary guidelines for treating patients with hypercholesterolemia or MS must be carefully planned in such a way that the incorporation of ChO into the diet should be controlled and nutritional background be considered.

Keywords: chia oil; endothelial function; glucose intolerance; hypercholesterolemia; metabolic syndrome; rabbits

Citation: Jerez, S.; Medina, A.; Alarcón, G.; Sierra, L.; Medina, M. Chia Seed Oil Intake: Is It Beneficial for Preventing Cardiovascular Risk Factors? *Biol. Life Sci. Forum* **2021**, *8*, 7. https://doi.org/10.3390/blsf2021008007

Academic Editors: Loreto Muñoz and Claudia M. Haros

Published: 28 December 2021

Publisher's Note: MDPI stays neutral with regard to jurisdictional claims in published maps and institutional affiliations.

1. Introduction

A cardiovascular risk factor is defined as a measurable characteristic that is causally associated with increased cardiovascular disease (CVD) frequency and that is a significant independent predictor of an increased risk of presenting with the disease. Traditional CVD risk factors are modifiable and non-modifiable. Modifiable CVD risk factors are those that can be reduced or controlled with altered behavior: tobacco use, hypertension, diabetes mellitus, hypercholesterolemia, physical inactivity and obesity. Diet is a key modifiable risk factor in the prevention and risk reduction of CVD. Nutritional interventions include ω-3 fatty acids, which were associated with reduced cardiovascular mortality and other cardiovascular outcomes.

Chia seeds have oil contents ranging from 25% to 40%, in which 60% is n-3 alpha-linolenic acid (18:3, ALA) and 20% is n-6 linoleic acid (18:2, LA) [1]. Of all the known food sources, chia contains the highest concentration of these fatty acids. The beneficial effects of chia seed on risk factors for CVD have been widely studied [2]. However, there is little evidence regarding the effects of chia seed oil (ChO). Han et al. [3] reported that ChO prevents high-fat diet-induced hyperlipidemia and oxidative stress in mice. Recently, Enes et al. [4] found that ChO is able to improve glucose tolerance and restore the energy fuel system in the livers of rats fed a high-fat diet. Furthermore, ChO supplementation changes body composition and activates the insulin signaling cascade in the skeletal muscle tissue of obese animals [5].

Endothelial dysfunction marks a stage of atherosclerosis and is an important prognostic marker for CVD. It has been reported that endothelial functionality is positively related to the proportion of ALA in plasma. Furthermore, several studies demonstrate that ALA is an endothelial protective factor [6]. In this sense, we demonstrate the beneficial effects of dietary intervention with ChO on endothelial function by using a rabbit model of hypercholesterolemia [7] and metabolic syndrome (a cluster of CVD risk factors including obesity, excessive visceral fat storage, dyslipidemia, hypertension and insulin resistance). However, some undesirable side effects on insulin resistance were found [8].

Thus, the aim of the present work was to discuss the effects of dietary interventions with ChO on animal models of CVD risk factors induced by high cholesterol or high fat diets.

2. Materials and Methods

2.1. Animals and Diets

Male hybrid rabbits initially weighing 850–1000 g were housed individually in gridded cages on a constant 12-h light/dark cycle under controlled temperature and conditions. After one week acclimation period, animals were randomized and separated in groups (n = 8 each): fed on regular rabbit chow (CD); fed on CD supplemented with either 10% (CD-ChO^{10}) or 3% of ChO (CD-ChO^{3}); fed on CD supplemented with 224.10 g/kg of chia seed (CD-Ch); fed on CD supplemented with 1% cholesterol (HD); fed on CD supplemented with 8% lard and 10% corn oil (HFD); fed on HD supplemented with 10% ChO (HD-ChO); and fed on HFD in which 3% of the oil source (corn oil) was replaced with ChO (HFD-ChO). Details of the diets are shown in Table 1. Rabbits were fed 180 g of the appropriate dietary treatment per day for 6 weeks. Daily energy intake from fat was similar in HFD and HFD-Ch. The dose of ALA was of 4.9 g (CD-ChO^{10} and CD-Ch) and 3 g/day (CD-ChO^{3}), all in agreement with the American Heart Association recommendations.

Table 1. Composition of experimental diets (g/100 g diet).

	CD	CD-ChO^{10}	CD-ChO^{3}	CD-Ch	HD	HD-ChO^{10}	HFD	HFD-ChO^{3}
Carbo hydrate	34 ± 2	34 ± 2	34 ± 2	43.4 ± 6	34 ± 2	34 ± 2	34 ± 2	34 ± 2
Protein	15 ± 3	15 ± 3	15 ± 3	18.7 ± 1	15 ± 3	15 ± 3	15 ± 3	15 ± 3
Fiber	15 ± 2	15 ± 2	15 ± 2	23 ± 2	15 ± 2	15 ± 2	15 ± 2	15 ± 2
Total fat	3.1 ± 0.1	13.1 ± 1.6	6.2 ±0.3	9.9 ± 0.2	3.0 ± 0.2	13.2 ± 1.4	21 ± 1.7	21 ± 1.4
Fatty acids								
Total SFA	21.4 ± 0.1	16.1 ± 0.9	16.9 ± 0.2	-	19.2 ± 0.6	18.1 ± 0.2	23.9 ± 0.9	22.6 ± 0.2
Total MUFA	17.6 ± 0.1	21.1 ± 0.5	20.0 ± 0.3	-	15.6 ± 0.1	15.2 ± 0.3	22.4 ± 0.5	22.3 ± 0.5
Total PUFA	56.7 ± 0.5	64.3 ± 0.4	64.0 ± 0.4	-	66.6 ± 0.9	66.7 ± 0.5	53.9 ± 0.1	58.7 ± 0.5
Total calories (Kcal)	250	332	275	359	250	332	410	411

SFA: saturated fatty acid; MUFA: monounsaturated fatty acids; PUFA: polyunsaturated fatty acids. CD: rabbits fed a control diet CD; CD-ChO10: fed a CD supplemented with 10% chia oil; CD-ChO^{3}: fed a CD supplemented with 3% chia oil; CD-Ch: fed a CD supplemented with 224 g/kg chia seed; HD: fed a CD supplemented with 1% cholesterol; CD-ChO^{10}: fed a HD supplemented with 10% chia oil; HFD: fed a CD supplemented with 10% corn oil and 8% lard; HFD-ChO3: fed a CD supplemented with 3% chia oil, 7% corn oil and 8% lard.

2.2. Clinical and Biochemical Parameters

An intraperitoneal glucose tolerance test (GTT) was performed two days before the end of the 6 weeks of feeding [8]. At the end of the dietary intervention, food was withdrawn for 12 h and animals were anesthetized. Mean arterial blood pressure (MAP) and heart rate (HR) were measured directly in the carotid artery through a catheter connected to a pressure transducer (Gould-Statham P23, Oxnard, CA, USA) and recorded using a data acquisition system (Biopac MP100, Aero Camino, Goleta, CA, USA). After MAP measurement, blood samples were collected through the catheter inserted in the carotid artery. Then, using surgical techniques, a midline incision was made in the rabbit and adipose tissues from the abdominal areas were collected and weighed. The visceral abdominal fat (VAF) was expressed as a percentage of the total body weight: (fat weight/animal weight) × 100. Plasma total cholesterol (TC), HDL-C, low density lipoprotein (LDL-C), TG, and FG were measured by using colorimetric reactions with commercial kits (Wiener, Rosario, Argentina).

The triglyceride-glucose index (TyG index) is a simple marker that has a high correlation with the degree of insulin resistance measured by hyperinsulinemic-euglycemic clamp studies. The TyG index was calculated as ln (fasting triglycerides ($mg{\cdot}dL^{-1}$) × fasting glucose ($mg{\cdot}dL^{-1}$).

Fatty acids from plasma were measured by gas chromatography according to previous work [8].

2.3. Vascular Function Assessment

Isometric contractions from aortic rings were measured by using force-displacement transducers (Grass Technologies, West Warwick, RI, USA) and were recorded under an initial tension of 2 g, which had been found to be the optimal tension for KCl-induced contraction (96 mM). The endothelial function was evaluated by testing the relaxation induced by a concentration response curve (CRC) to acetylcholine (Ach, 10^{-8}–5×10^{-6} M) in aortic rings pre contracted with phenylephrine (Phe) 5×10^{-6} M.

2.4. Statistical Analyses

CRC were fitted using a nonlinear interactive fitting program (GraphPad Prism 3.0; GraphPad Software Inc., San Diego, CA, USA). Agonist potencies were calculated as pEC50 (negative logarithm of the molar concentration of agonist producing 50% of the maximum response), and maximum response was expressed as Rmax (maximum effect elicited by the agonist). Investigators were blinded to treatment until data analysis. Results are reported as mean ± standard error of the mean (SEM).

The Shapiro–Wilks goodness-of-fit test was used to test for normal distribution. Statistically significant differences were calculated by one- or two-way analysis of variance (followed by Duncan's post-test) or unpaired Student's *t*-test; $p < 0.05$ was considered statistically significant.

3. Results and Discussion

3.1. Clinical and Biochemical Parameters

Independent of the differences in caloric intake, animal body weights did not differ significantly among the groups (Table 2). As was previously reported [2], TC, TG, MAP and TyG index were higher in rabbits fed a HD than CD group, and ChO 10% supplementation normalized TG and MAP. Unexpectedly, the control group, CD-ChO^{10}, showed glucose intolerance and higher levels of FG and TyG index compared with the CD group. These results disagree with those of other authors [4,5]. Several studies find that dietary supplement with chia seed improves glucose tolerance and insulin sensitivity. Chicco et al. [9] report that dietary chia seed improves adiposity and normalizes hypertriglyceridemia and insulin resistance in an experimental model of dyslipidemia and insulin resistance. Poudyal et al. [10] show that a chia seed-supplemented high carbohydrate-high fat diet does not reduce total body fat but induces lipid redistribution away from the abdominal area and improves glucose tolerance and insulin sensitivity. However, all these studies were performed with the chia seed, the composition of which is different than ChO [1]. Chia seeds contain high levels of fiber, and this would be an advantage with respect to ChO. Thus, in a recent study, we replaced the 10% of ChO by chia seed in such a way that the contents of ALA is equivalent. An increase in VAF and glucose intolerance were found in this group. We hypothesized that ALA intake was too high and would be responsible for such alterations. Thus, ChO intake was reduced (4.9 g to 3 g per day) and no differences were found in the CD-ChO^{3} group as compared with the CD group. Therefore, we aimed to study the effects of dietary intervention with 3% ChO on a rabbit model of metabolic syndrome induced by HFD. According to Alarcon et al. [8], TG, FG, VAF, glucose tolerance test and the n-6/n-3 fatty acid ratio were higher and HDL-C was lower in the HFD than in the CD group (Table 2). Although the replacement of 3% corn oil with ChO into the HFD lowered the TG levels and the n-6/n-3 fatty acid ratio, it failed to improve HDL-C levels. In addition, ChO worsened the deleterious effects of the HFD on VAF, FG and glucose intolerance.

Table 2. Clinical and biochemical parameters from rabbits fed a control diet (CD); fed a CD supplemented with 10% chia oil (CD-ChO^{10}); fed a CD supplemented with 3% chia oil (CD-ChO^{3}); fed a CD supplemented with 224 g/kg chia seed (CD-Ch); fed a CD supplemented with 1% cholesterol (HD); fed a HD supplemented with 10% chia oil (CD-ChO^{10}); fed a CD supplemented with 10% corn oil and 8% lard (HFD); and fed a CD supplemented with 3% chia oil, 7% corn oil and 8% lard (HFD-ChO^{3}).

	CD	CD-ChO^{10}	CD-ChO^{3}	CD-Ch	HD	HD-ChO^{10}	HFD	HFD-ChO^{3}
BW	1.9 ± 0.1	1.9 ± 0.6	1.8 ± 0.6	2.0 ± 0.6	2.1 ± 0.4	1.8 ± 0.1	2.0 ± 0.4	1.9 ± 0.2
VAF	1.0 ± 0.2	1.3 ± 0.2	1.4 ± 0.4	2.1 ± 0.1 *	0.70 ± 0.3	0.93 ± 0.2	2.3 ± 0.1 *	3.4 ± 0.3 *
FG	113 ± 3	132 ± 6 *	105 ± 23	109 ± 11	115 ± 4	139 ± 8 *	126 ± 6 *	148 ± 11 *
Glu-60	183 ± 5	200 ± 16	170 ± 35	205 ± 27	194 ± 10	210 ± 14	248 ± 17 *	277 ± 14*
Glu-120	138 ± 3	161 ± 9 *	140 ± 19	160 ± 19 *	138 ± 6	171 ± 11 *	172 ± 9 *	241 ± 21 *
TC	59 ± 60	53 ± 16	49.8 ± 12	95.5 ± 38	872 ± 114 *	783 ± 278 *	78 ± 5	81 ± 5
LDL-C	29 ± 8	24.6 ± 4.0	23.5 ± 3.8	24 ± 1.4	666 ± 99 *	705 ± 291	47 ± 7	42 ± 7
HDL-C	51 ± 7	25 ± 4 *	24 ± 2 *	51 ± 2	164 ± 45 *	128 ± 47 *	24 ±3 *	20 ± 0.4 *
TG	113 ± 14	120 ± 29	122 ± 15	65 ± 2.4	222 ± 33 *	102 ± 2	192 ± 22 *	104 ± 9
TyG	8.3 ± 0.2	9.5 ± 0.1 *	8.6 ± 0.2		10.2 ± 0.5 *	9.3 ± 0.2 *	9.3 ± 0.3 *	8.9 ± 0.1 *
MAP	56.0 ± 2.6	61 ± 6	44.6 ± 4.6	48.3 ± 2.4	73 ± 2 *	64.8 ± 4.6	57 ± 5	45 ± 2
HR	265 ± 25	224 ± 5	249 ± 27	253 ± 5	226 ± 11	253 ± 7	282 ± 27	325 ± 44
n-6/n-3	9.2 ± 0.2	4.0 ± 0.3	4.2 ± 0.1	-	7.6 ± 0.2	3.4 ± 0.1	35 ± 0.2 *	12 ± 1.3

BW: body weight (kg); VAF: visceral abdominal fat (%); FG: fasting glucose (mg/dL); Glu-60: glucose level 60 min post intraperitoneal injection of glucose (mg/dL); Glu-120: glucose level 120 min post intraperitoneal injection of glucose (mg/dL); TC: total cholesterol (mg/dL); TG: triglycerides (mg/dL); MAP: mean arterial blood pressure (mmHg); HR: heart rate (bpm); * statistically different from CD.

3.2. Vascular Function

Relaxation to Ach in blood vessels is a validated marker of endothelial function. Endothelium-dependent relaxation to Ach was lower in aortic rings from the HD (37 ± 4%) and HFD (44 ± 6%) group than those from the CD group (68 ± 9%). Dietary intervention with ChO partially normalized Ach relaxation in the HD-ChO group (51 ± 5%) and significantly improved endothelial function from HFD-ChO (60 ± 4%). These results confirm the beneficial effect of ALA on endothelial function.

4. Conclusions

Many authors claim the benefit of chia seed on CVD risk factors. However, the effects of dietary interventions with ChO have been little explored. Results from our studies show that incorporation of ChO into the diet has beneficial and harmful effects. The beneficial effects include the reduction of TG levels and MAP and the improvement of endothelial function. However, ChO may induce glucose intolerance, and even more importantly, it may worsen the HFD-induced deleterious effects on VAF and FG. The experimental animal models were previously characterized, and they mimic the CVD risk factors in human beings. Therefore, our results support the view that dietary guidelines for treating patients with CVD risk factors must be carefully planned in such a way that before the incorporation of ChO into the diet, the nutritional background should be considered.

Institutional Review Board Statement: All animal care and use programs were performed according to the Guide for the Care and Use of Laboratory Animals (NIH Publication 86 to 23, revised 1985), and approved by the Institutional Animal Care and Use Committee (CICUAL-UNT; protocol code 021/2019).

Acknowledgments: This work was supported by grant Ia ValSe-Food-CYTED (119RT0567) and Préstamo BID PICT 2015-1164, and PIUNT 2018 G621.

References

1. Ixtaina, V.Y.; Martinez, M.L.; Spotorno, V.; Mateo, C.M.; Maestri, D.M.; Diehl, B.W.; Nolasco, S.M.; Tomas, M.C. Characterization of chia seed oils obtained by pressing and solvent extraction. *J. Food Anal.* **2011**, *24*, 166–174. [CrossRef]
2. Parker, J.; Schellenberger, A.N.; Roe, A.L.; Oketch-Rabah, H.; Calderon, A.I. Therapeutic perspectives on chia seed and its oil: A review. *Planta Med.* **2018**, *84*, 606–612. [CrossRef] [PubMed]
3. Han, K.; Li, X.; Zhang, Y.; He, Y.; Hu, R.; Lu, X.; Li, Q.; Hui, J. Chia seed oil prevents high fat diet induced hyperlipidemia and oxidative stress in mice. *Eur. J. Lipid Sci. Technol.* **2020**, *122*, 1900443. [CrossRef]
4. Enes, B.N.; Moreira, L.P.D.; Toledo, R.C.L.; Moraes, É.A.; Moreira, M.; Hermsdorff, H.H.; Noratto, G.; Mertens-Talcott, S.; Talcott, S.; Martino, H. Effect of different fractions of chia (*Salvia hispanica* L.) on glucose metabolism, in vivo and in vitro. *J. Funct. Foods* **2020**, *71*, 104026. [CrossRef]
5. Fonte-Faria, T.; Citelli, M.; Atella, G.C.; Raposo, H.F.; Zago, L.; de Souza, T.; da Silva, S.; Barja-Fidalgo, C. Chia oil supplementation changes body composition and activates insulin signaling cascade in skeletal muscle tissue of obese animals. *Nutrition* **2019**, *58*, 167–174. [CrossRef] [PubMed]
6. Steer, P.; Vessby, B.; Lind, L. Endothelial vasodilatory function is related to the proportions of saturated fatty acids and alpha-linolenic acid in young men, but not in women. *Eur. J. Clin. Investig.* **2003**, *33*, 390–396. [CrossRef] [PubMed]
7. Sierra, L.; Roco, J.; Alarcon, G.; Medina, M.; Van Nieuwenhove, C.; Peral, M.; Jerez, S. Dietary intervention with *Salvia hispanica* (Chia) oil improves vascular function in rabbits under hypercholesterolaemic conditions. *J. Funct. Foods.* **2015**, *14*, 641–649. [CrossRef]
8. Alarcon, G.; Medina, A.; Martin Alzogaray, F.; Sierra, L.; Roco, J.; Van Nieuwenhove, C.; Medina, M.; Jerez, S. Partial replacement of corn oil with chia oil into a high fat diet produces either beneficial and deleterious effects on metabolic and vascular alterations in rabbits. *Pharmanutrition* **2020**, *14*, 110218. [CrossRef]
9. Chicco, A.G.; D'Alessandro, M.E.; Hein, G.J.; Oliva, M.E.; Lombardo, Y.B. Dietary chia seed (*Salvia hispanica* L.) rich in alpha-linolenic acid improves adiposity and normalises hypertriacylglycerolaemia and insulin resistance in dyslipaemic rats. *Br. J. Nutr.* **2009**, *101*, 41–50. [CrossRef]
10. Poudyal, H.; Panchal, S.; Ward, L.; Waanders, J.; Brown, L. Lipid redistribution by α-linolenic acid-rich chia seed inhibits stearoyl-CoA desaturase-1 and induces cardiac and hepatic protection in diet-induced obese rats. *J. Nutr. Biochem.* **2003**, *24*, 1041–1052. [CrossRef]

biology and life sciences forum

MDPI

Proceeding Paper

Development of Jams with Ancestral Seed Aggregates †

Bailey Jannika, Mezzatesta Pablo, Farah Silvia, Figueras Tatiana and Raimondo Emilia *

Universidad Juan Agustín Maza, Mendoza CP5519, Argentina; jannikabailey@gmail.com (B.J.); pablomezzatesta@gmail.com (M.P.); farahsilvia1@hotmail.com (F.S.); tatianafigueras@gmail.com (F.T.)
* Correspondence: emilia.raimondo@gmail.com
† Presented at the III Conference la ValSe-Food and VI Symposium Chia-Link Network, online, 15–17 November 2021.

Abstract: Small-scale food producers have been negatively impacted by the present pandemic and have been forced to use innovations with low-risk products as a means to increase sales. The object was to determine variations in the nutrient profile of peach jam with the introduction of amaranth or quinoa seeds, the latter having been rinsed beforehand to reduce saponin content. Three varieties of the jam were made, and these were subjected to a sensory evaluation by a panel of 30 untrained judges (consumers) and analysed to determine the variation in their composition as a result of the addition of the seeds. To the basic preparation, which consisted of peaches and sugar (PJ), 20% of quinoa seeds were added (QJ) at the bottling stage. To the third jam preparation, amaranth seeds were added in the same proportion (AJ). Official analytical techniques were used to determine their nutrient profile. The protein content increased from 0.23 g% (PJ) to 2.52 g% (QJ) and 3.38 g% (AJ). Total fat increased from 0.35 g% (PJ) to 0.74 g% (QJ) and 1.72 g% (AJ). Fibre increased from 2.13 g% (PJ) to 4.24 g% (QJ) and 2.86 g% (AJ). The incorporation of amaranth and quinoa improved protein profile, fibre and total fat intake and also resulted in a jam with a better nutrient profile, although there was only a slight reduction in carbohydrates, from 68 g% to 66 g%, after the seeds were added. Plum and apricot jam were also tested, and in all instances, the results were similar.

Keywords: amaranth; fruit jam; nutritional profile; vegetable protein; quinoa

Citation: Jannika, B.; Pablo, M.; Silvia, F.; Tatiana, F.; Emilia, R. Development of Jams with Ancestral Seed Aggregates. *Biol. Life Sci. Forum* **2021**, *8*, 9. https://doi.org/10.3390/blsf2021008009

Academic Editors: Loreto Muñoz and Claudia M. Haros

Published: 15 February 2022

1. Introduction

The current COVID-19 pandemic led to a drop in sales in all sectors, especially affecting small food processors, which is why innovating in low-risk products allows them to expand their sales [1]. Small producers do not have adequate infrastructures to produce foods that could put the consumer's health at risk, such as highly acidic preserves, due to the possibility of the development of botulism [2]. For this reason, they are encouraged to produce low-risk foods, such as jams from different fruits. To be considered a jam, its consistency must be spreadable, have a soluble solids content of 65 °Brix, be free of skin, seeds or pits from the fruit, except those such as strawberries that cannot have their skin or seeds removed [3]. Their spreadability makes them very popular because they are eaten together with biscuits or bread at breakfast or snacks. One of the problems with these jams is their low nutritional value, as they only provide simple sugars and energy. Improving the nutrient profile by adding ancestral seeds not only increases the range of products offered by these small entrepreneurs but also improves the nutritional value of the jams, with the consequent impact on consumers' health [4].

The ancestral seeds chosen were (*Amaranthus caudatus*) and quinoa (*Chenopodium quinoa Willd.*), both with very good crop yields in Mendoza, Argentina, and especially in the northwest of the country. Amaranth was chosen for its protein content of 16 to 18 g% [5], with a complete amino acid profile, resulting in a protein of high biological value. Within its lipid fraction, it provides squalene, which is related to lowering cholesterol, according to a study published by Gonor in 2006 [6].

Quinoa (*Chenopodium quinoa Willd.*) [7] is an ancient crop that is ideal for populations suffering from protein malnutrition. Depending on the variety, its content varies from 13 to 21%, with a complete amino acid profile. According to FAO (2013) [7], it is a food that contributes to global food security. It has a dietary fibre content of around 12 g%, which reduces the digestibility of the grain, but provides satiety. It has a high vitamin E content, which acts as an antioxidant, protecting the fatty acids it provides [8], of which 50% correspond to polyunsaturated fatty acids (omega 6) and 25% to monounsaturated fatty acids (omega 9) [9].

The objective was therefore to determine the variation in the nutritional profile of a peach jam by adding amaranth seeds or quinoa seeds. This experiment was then repeated with plum and apricot jam.

2. Materials and Methods

Peach jam (PJ—base jam), peach jam with quinoa seeds (QJ—quinoa jam) and peach jam with amaranth seeds (AJ—amaranth jam) were made in triplicate. Plum jam, with and without seeds, and apricot jam, with and without seeds, were produced with the same characteristics.

2.1. Preparation Methods

The peach jam was made in triplicate. This was done on a pilot scale, using an open pan with permanent agitation. The peaches, previously washed, peeled and pitted, were ground to obtain the pulp, to which 300 mg of citric acid per kg was added to avoid oxidation. The ratio was 700 g of sugar per kg of pulp. The pulp and only 10% of the calculated sugar were placed in the pan and boiled with constant stirring. After approximately 30 min, when the pulp reached 35–40 °Brix, the rest of the sugar was added, reaching the final point of 65 °Brix after approximately 45 min. This results in a light-coloured, peach-scented jam. This was repeated using plums as a base and then apricots, with the only difference being that the latter was not peeled but was only pitted and ground.

When working with quinoa, the seed was first washed seven times with twice the volume of water until no more foam was formed. This removes all the saponins. If this operation is not carried out correctly, the seed takes on a bitter taste, which is transferred to the jam. Once the seed has been washed and drained, it is added in a proportion of 20% of the pulp when the rest of the sugar is added, which is when there are only a few minutes of cooking time left.

In the case of the amaranth seed, it was washed once, drained and added in the same proportion and at the same time as detailed for quinoa.

The same steps were followed for plum seed jam or apricot seed jam.

2.2. Laboratory Analysis

The following methods were used to determine the nutritional composition of the different jams, with and without seeds [10]:

Humidity: Method of A.O.A.C 950.46 B. Indirect method by drying in an oven at 100–105 °C, until constant weight is achieved.

Total fat: Direct method by extraction with ethyl ether (crude fat), Soxhlet gravimetric method (A.O.A.C. 960.39, 1990) was used.

Fibres: Acid alkaline attack (AOAC, 15th edition 1990) was used.

Crude protein: Kjeldahl method, (A.O.A.C. 928.08, 1990), determining nitrogen, using 6.25 as a protein conversion factor.

Ashes: Direct Method (A.O.A.C. 923.03, 1990): by incineration in muffle (at 500 ± 10 °C), until constant ash weight.

Carbohydrates: determined by difference, by the following formula:

100 − (weight in grams [protein + fat + water + ash + fibres]), in 100 g of food.

Energy value: by calculation

Energy value (kcal) = By calculation. The conversion is 2000 kcal = 8400 kJ.

Statistical analysis. The Kolmogorov–Smirnova test was applied to check the assumption of normality of the residuals, and the Levene's test for homoscedasticity of variances was applied to check the homoscedasticity of the residuals. To compare the means of nutrients between the different jams, an ANOVA test was applied, and for post hoc analysis, Tukey's test was used. The analysis was carried out with the SPSS® statistical package.

2.3. Sensory Analysis

In order to assess the acceptability of the jams, an acceptance test was carried out with 30 untrained judges.

3. Results

3.1. Peach Marmalade with and without Added Seeds

Table 1 shows the nutritional composition of the peach jam (PJ) compared to the amaranth seed jam (AJ) and the quinoa jam (QJ).

Table 1. Peach Jam composition.

	PJ	AJ	QJ
Protein g%	0.23 ± 0.02	3.38 ± 0.03	2.74 ± 0.02
Carbohydrates	68.00 ± 0.19	66.36 ± 0.20	66.27 ± 0.32
Total Fats g%	0.35 ± 0.02	1.72 ± 0.03	0.74 ± 0.03
Saturadted fats g%	0.02 ± 0.01	0.26 ± 0.02	0.07 ± 0.02
Ashes g%	0.65 ± 0.02	0.64 ± 0.04	0.97 ± 0.02
Humidity g%	28.64 ± 0.22	25.03 ± 0.15	28.64 ± 0.22
Dietary fibre g%	2.13 ± 0.09	2.86 ± 0.03	4.24 ± 0.09
Sodium mg%	17 ± 1	14 ± 3.61	17 ± 1
Energy value kcal	276 ± 0.82	294 ± 0.57	283 ± 1.24
Energy value kJ	1159 ± 3.43	1237 ± 2.40	1187 ± 5.19

Protein content increased from 0.23 g% (PJ) to 2.52 g% (QJ) and 3.38 g% (AJ). Total fat increased from 0.35 g% (PJ) to 0.74 g% (QJ) and 1.72 g% (AJ). Fibre increased from 2.13 g% (PJ) to 4.24 g% (QJ) and 2.86 g% (AJ). With the incorporation of amaranth or quinoa, the contribution of protein, fibre and total fat was improved, achieving a jam with a better nutritional profile, the decrease in carbohydrates not being significant, going from 68 g% to 66 g%, due to the incorporation of the seeds.

Applying statistics to the values obtained, it turns out that the analysis of the assumptions of normality and homogeneity of the residues is fulfilled with a p-value greater than 0.05. For ANOVA tests comparing peach jam (PJ), with added amaranth (AJ) and with added quinoa (QJ), statistically significant differences were found in the following nutrients ($p < 0.001$): protein, carbohydrates, dietary fibre, total fat and energy content.

3.2. Plum Jam with and without Added Seeds

Table 2 shows the nutritional composition of the plum jam (PlJ) compared to the jam with amaranth seeds (APlJ) and the jam with quinoa (QPlJ).

Again, the highest protein and lipid values were obtained with the addition of amaranth seeds, together with lower carbohydrate content and intermediate fibre content. The highest fibre content is obtained in the quinoa seed jam.

For the ANOVA tests comparing plum jam (PlJ), with the addition of amaranth (APlJ) and with the addition of quinoa (QPlJ), statistically significant differences were found in the following nutrients ($p < 0.001$): protein, carbohydrates, dietary fibre, total fat and energy content.

Table 2. Plum Jam composition.

	PlJ	APlJ	QPlJ
Protein g%	0.38 ± 0.03	3.50 ± 0.02	2.86 ± 0.02
Carbohydrates	65.33 ± 0.31	64.22 ± 0.15	64.14 ± 0.31
Total Fats g%	0.33 ± 0.02	1.70 ± 0.02	0.73 ± 0.04
Saturadted fats g%	0.03 ± 0.01	0.26 ± 0.02	0.08 ± 0.02
Ashes g%	0.51 ± 0.06	0.52 ± 0.03	0.85 ± 0.02
Humidity g%	29.28 ± 0.33	25.56 ± 0.04	25.55 ± 0.30
Dietary fibre g%	4.17 ± 0.14	4.50 ± 0.06	5.87 ± 0.01
Sodium mg%	15.33 ± 2.08	11 ± 2	15 ± 2
Energy value kcal	266 ± 1.41	286 ± 0.37	275 ± 1.35
Energy value kJ	1116 ± 5.93	1202 ± 1.55	1153 ± 5.68

3.3. *Apricot Jam with and without Added Seeds*

Table 3 shows the nutritional composition of the apricot jam (ApJ) compared to the amaranth seed jam (AApJ) and the quinoa jam (QApJ).

Table 3. Apricot Jam composition.

	ApJ	AApJ	QApJ
Protein g%	0.20 ± 0.03	3.36 ± 0.03	2.84 ± 0.10
Carbohydrates	66.50 ± 1.15	65.16 ± 0.13	64.59 ± 0.14
Total Fats g%	0.20 ± 0.03	1.60 ± 0.04	0.73 ± 0.02
Saturadted fats g%	0.02 ± 0.01	0.26 ± 0.02	0.07 ± 0.02
Ashes g%	0.63 ± 0.03	0.61 ± 0.01	0.88 ± 0.04
Humidity g%	29.71 ± 1.23	25.90 ± 0.08	25.43 ± 0.27
Dietary fibre g%	2.76 ± 0.06	3.37 ± 0.01	5.53 ± 0.09
Sodium mg%	7 ± 2	11 ± 1	16 ± 2
Energy value kcal	269 ± 4.47	288 ± 0.25	276 ± 0.68
Energy value kJ	1128 ± 18.77	1212 ± 1.06	1160 ± 2.86

The observations for peach and plum are repeated.

4. Discussion

The addition of amaranth and quinoa improved the protein content of high biological value provided by the seeds, considering that the jams do not provide protein or lipids. Fats are also provided by quinoa and amaranth, with the advantages of the healthy fatty acid profile of these seeds. With regard to fibre, fibre consumption increases, with all the benefits that the intake of this nutrient brings.

5. Conclusions

The addition of amaranth and quinoa seeds to different jams improves their nutritional profile in all cases. It is a simple practice to apply in small producers, increasing the supply of healthier foods and revaluing ancestral seeds.

Author Contributions: All authors contributed to this proceeding paper article. All authors have read and agreed to the published version of the manuscript.

Funding: This research received no external funding.

Institutional Review Board Statement: Not applicable.

Informed Consent Statement: Not applicable.

Data Availability Statement: Not applicable.

Acknowledgments: This work was supported by a grant from La ValSe-Food-CYTED (119RT0567) and Universidad Juan Agustin Maza.

Conflicts of Interest: The authors declare no conflict of interest.

References

1. Bullemore-Campbell, J.; Cristóbal-Fransi, E. Commercial address in pandemic time: El impacto del COVID-19 en la gestión de ventas. *Inf. Tecnol.* **2021**, *32*, 199–208. [CrossRef]
2. Centro para el Control y la Prevención de Enfermedades (CDC). Los Alimentos en Conserva Envasados en la casa y el Botulismo. 2021. Available online: https://www.cdc.gov/foodsafety/es/communication/home-canning-and-botulism.html (accessed on 14 June 2021).
3. ANMAT Código Alimentario Argentino. Available online: https://www.argentina.gob.ar/anmat/codigoalimentario (accessed on 12 June 2021).
4. National Heart Foundation of Australia. Rapid Review of the Evidence. Effectiveness of Food Reformulation as a Strategy to Improve Population Health. 2012. Available online: https://www.heartfoundation.org.au/getmedia/2bcb1637-1a86-48fc-bb6c-b243c4746272/RapidReview_FoodReformulation.pdf (accessed on 14 July 2021).
5. Algara Suárez Paola, Gallegos Martínez Josefina, Reyes Hernández Jaime. Amaranto: Efectos en la Nutrición y la Salud. Revista Académica de Investigación. Editada por Eumed.net, España. 2013. Available online: https://dialnet.unirioja.es/servlet/articulo?codigo=7324335 (accessed on 20 July 2021).
6. Gonor, K.V.; Pogozheva, A.V.; Derbeneva, S.A.; Mal'tsev GIu Trushina, E.N.; Mustafina, O.K. The influence of a diet with including amaranth oil on antioxidant and immune status in patients with ischemic heart disease and hyperlipoproteidemia. *Vopr. Pitan.* **2006**, *75*, 30–33. [PubMed]
7. FAO. La Quinoa: Cultivo Milenario Para Contribuir a la Seguridad Alimentaria Mundial. 2011. Available online: http://www.fao.org/3/aq287s/aq287s.pdf (accessed on 22 April 2021).
8. Ng, S.C.; Anderson, A.; Coker, J.; Ondrus, M. Characterization of lipid oxidation products in quinoa (Chenopodium quinoa). *Food Chem.* **2007**, *101*, 185 192. [CrossRef]
9. Sánchez-Urdaneta, A.B.; Montero-Quintero, K.C.; González-Redondo, P.; Molina, E.; Bracho-Bravo, B.; Moreno-Rojas, R. Hypolipidemic and Hypoglycaemic Effect of Wholemeal Bread with Amaranth (Amaranthus dubius Mart. ex Thell.) on Sprague Dawley Rats. *Foods* **2020**, *9*, 707. [CrossRef] [PubMed]
10. AOAC Official Methods of Analysis. 1990. Available online: https://archive.org/stream/gov.law.aoac.methods.1.1990/aoac.methods.1.1990_djvu.txt (accessed on 13 April 2021).

biology and life sciences forum

MDPI

Proceeding Paper

Chia Oil Microcapsules Obtained by Different Drying Methods †

Claudia N. Copado [1], Vanesa Y. Ixtaina [1,2] and Mabel C. Tomás [1,*]

1 Centro de Investigación y Desarrollo en Criotecnología de Alimentos (CIDCA), CCT La Plata (CONICET), Facultad de Ciencias Exactas (FCE), Universidad Nacional de La Plata (UNLP), 47 y 116, La Plata 1900, Argentina; claudiacopado16@gmail.com (C.N.C.); vanesaix@hotmail.com (V.Y.I.)

2 Facultad de Ciencias Agrarias y Forestales (FCAyF, UNLP), 60 y 119, La Plata 1900, Argentina

* Correspondence: mabtom@hotmail.com

† Presented at the III Conference la ValSe-Food and VI Symposium Chia-Link Network, Online, 15–17 November 2021.

Abstract: A technology used to protect chia oil from lipid oxidation during processing and storage is microencapsulation. Thus, microcapsules containing chia oil could be applied as an ingredient to develop enriched foods with ω-3 fatty acids. The objective of this technology is to achieve high microencapsulation efficiency and provide greater oxidative stability to the chia oil. This work compares microcapsules obtained by different methods such as spray-drying and freeze-drying. To establish relationships between the microencapsulated chia oil using both methodologies and some of the characterization parameters studied, a multivariate analysis was carried out considering the microcapsules obtained from the parental emulsions with 10 or 15% *w*/*w* of chia oil, 10% *w*/*w* of lactose, and 10% *w*/*w* of sodium caseinate, whose aqueous phases were or not heat-treated at 60 or 100 °C, 30 min. The results show that the main components 1 (CP1) and 2 (CP2) explain 46.7 and 38.1% of the observed variability, respectively, totaling around 85%. The CP1 allowed separation of the microcapsules obtained by spray-drying from the freeze-drying ones, while the CP2 permitted to discriminate within the chia oil microencapsulated by freeze-drying, the systems whose aqueous phases were treated or not at 100 °C, 30 min from the rest of the microcapsules. The multivariate analysis made it possible to differentiate the microcapsules obtained by spray-drying and the freeze-drying ones. The former being associated with greater luminosity and microencapsulation efficiency, as well as a lower level of moisture content, water activity, and b* (blue-yellow component of the CIELab system) values.

Keywords: chia oil; freeze-dryer; microcapsules; spray-dryer

Citation: Copado, C.N.; Ixtaina, V.Y.; Tomás, M.C. Chia Oil Microcapsules Obtained by Different Drying Methods. *Biol. Life Sci. Forum* **2021**, *8*, 10. https://doi.org/10.3390/blsf2021008010

Academic Editors: Loreto Muñoz and Claudia M. Haros

Published: 25 January 2022

1. Introduction

The growing knowledge available about the relationship between food and health has prompted various actions in order to achieve an improvement in the composition of the diet through the incorporation of functional compounds. This reality has led the food industry to concentrate more and more on the development of functional foods [1].

The consumption of polyunsaturated fatty acids (PUFAs) ω-3 offers various health benefits; for example, protection against coronary heart disease, inflammatory disorders, asthma, and retinal diseases, as well as an improvement in brain function. Thus, the incorporation of these compounds into the diet is very important [2].

Oils from vegetable sources are good sources of essential fatty acids and other fatty acids derived from them. Regarding the lipid fraction of green leafy plants and some seeds, a relevant component is α-linolenic acid (ALA) [3]. Specifically, the oil content of chia seeds is about 25–40% *w*/*w*, with ALA (~65%) and linoleic acid (LA) (~20%) as the most abundant fatty acids, and a low percentage of saturated fatty acids (SFA) [4]. The high content of ω-3 fatty acids gives chia oil a high nutritional value. However, the unsaturation degree of PUFAs is responsible for the great susceptibility of this oil to lipid oxidation [5]. It is, therefore, essential to use some strategies such as microencapsulation to guard this oil during storage and/or processing. Among the different microencapsulation processes, spray-drying and freeze-drying have been applied to microencapsulate different types of oils. The features of the obtained microparticles depend on the microencapsulation process, the wall material, the wall/core ratio, and the characteristics of the parent emulsions before the drying process.

The Maillard reaction is a series of complex chemical reactions between the reducing end of a carbohydrate and the free amino group of a protein, accelerated under heat treatments [6]. Different studies found that Maillard reaction products (MRPs) offer the proteins a better emulsifying capability, foaming, heat stability, and solubility, than the native form. Additionally, MRPs have antioxidant activity, mainly due to the compounds formed from the Amadori rearrangement [6,7]. Thus, MRPs can be used as wall material to microencapsulate chia seed oil.

The objective of this study was to characterize the chia oil microcapsules obtained by spray drying and freeze-drying, utilizing MRPs generated from heat treatment of sodium caseinate and lactose as wall material, and to analyze the data by a multivariate statistical method verifying its ability to distinguish groups of microparticles.

2. Materials and Methods

2.1. Experimental Design and Preparation of Microcapsules

The experimental design proposed was selected for evaluating the influence of the oil content (10 or 15 g/100 g emulsion) and the aqueous phase heat treatment to stimulate the MRPs production. Different O/W emulsions were formulated with sodium caseinate (NaCas) (10% *w/w*), lactose (10% *w/w*) and two chia oil concentrations (10 and 15% *w/w*) (Table 1). The NaCas was mixed in distilled water (50 °C) under magnetic agitation. Afterward, the lactose was incorporated in the aqueous phase. The protein–carbohydrate mixture was heated (60 or 100 °C, 30 min) in a water bath. Nisine (0.0012 g/100 g) and potassium sorbate (0.1 g/100 g) were used for microbial growth inhibition. The emulsion was prepared in two stages. The pre-emulsification was carried out by Ultra Turrax T 25 equipment (IKA Labortechnik, Germany), using a S 25 N - 18 G dispersing tool (1 min, 9500 rpm). Then, the pre-emulsion was homogenized with a Panda 2K high-pressure valve homogenizer (GEA Niro Soavi, Parma, Italy) at 600 bars, for 4 cycles. The microcapsules were obtained by drying the emulsions in a Mini Spray Dryer B-290 (Büchi, Switzerland), using 170 °C for the inlet air and 75 °C for the outlet one, or in a freeze-dryer (L-A-B4-C, Rificor, Buenos Aires, Argentina). For this last process, the O/W emulsions (~100 g) were placed in 125 mm × 160 mm plastic trays, forming a ~1 cm thick layer and frozen at −20 ± 1 °C, 48 h, and then were transferred to −80 ± 1 °C for 24 h. Microcapsules were then obtained from the frozen emulsions by freeze-drying in laboratory-scale equipment using 1 atm of vacuum press for 48 h.

2.2. Moisture Content

This parameter was measured gravimetrically in a vacuum oven (Instrumentación Científica S.A., Buenos Aires, Argentina) according to Baik et al. (2004) [8].

2.3. Water Activity (a_w)

The a_w was determined using an AquaLab Water Activity Meter CX2 model (Decagon Devices Inc., Pullman, WA, USA) at 25 ± 0.5 °C.

Table 1. Experimental design of the microencapsulation of chia seed oil using two types of drying (spray-drying and freeze-drying).

Microencapsulation Process	System Code	% Oil (g/100 g of Parent Emulsion)	Oil Content (g/100 g of Microcapsules)	Lactose Content (%w/w)	NaCas Content (%w/w)	Heat Treatment
Spray-drying	STT-10 SD STT-15 SD	10 15	33.0 42.9	10	10	-
	TT60-10 SD TT60-15 SD	10 15	33.0 42.9	10	10	60 °C, 30 min
	TT100-10 SD TT100-15 SD	10 15	33.0 42.9	10	10	100 °C, 30 min
Freeze-drying	STT-10 FD STT-15 FD	10 15	33.0 42.9	10	10	-
	TT60-10 FD TT60-15 FD	10 15	33.0 42.9	10	10	60 °C, 30 min
	TT100-10 FD TT100-15 SD	10 15	33.0 42.9	10	10	100 °C, 30 min

NaCas, sodium caseinate.

2.4. Microencapsulation Efficiency (ME)

The free oil was quantified according to the method proposed by Augustin et al. [9] with some modifications. Approximately 1 g of powder was used, which was washed with hexane through a filter paper (Whatman N° 4). The difference of weight in the samples allowed calculation of the free oil concentration. The total oil of the microcapsules was considered equal to the initial oil. ME was calculated according to Equation (1).

$$EM(\%) = \frac{(\text{total oil} - \text{free oil})}{\text{total oil}} \times 100 \quad (1)$$

2.5. Color

The surface color of the microcapsules was measured using a Minolta color analyzer (CR-400, Konica Minolta Sensing Inc., Japan). This colorimeter was calibrated with a white standard, and color was evaluated using the CieLab system (L*, a*, and b*) [10].

2.6. Statistical Analysis

A multivariate analysis (PCA principal component analysis) was performed considering the microcapsules obtained by both drying methods from systems with the same formulation. This analysis was carried out using the Software MultiVariateStatistical Package (MVSP) version 3.1 (Kovach Computing Services, Wales, UK).

3. Results and Discussion

In order to establish relationships between the microencapsulated chia oil using both methodologies and some of the characterization parameters studied, a multivariate analysis was carried out considering the microcapsules obtained from the parental emulsions with 10 or 15% w/w of chia oil, 10% w/w of lactose whose aqueous phases were or not thermally treated at 60 or 100 °C, 30 min. The results obtained for each parameter studied are shown in the Tables 2 and 3, and Figure 1 represents the result of principal component analysis (PCA).

Table 2. Parameters of the microcapsules obtained by spray-drying.

Microcapsules	H (% d.b.)	a_w (25 °C)	ME (%)	Color		
				L*	a*	b*
STT-10	1.34	0.52	83.96	93.52	−1.01	12.81
STT-15	0.74	0.48	74.67	92.21	−1.01	14.06
TT60-10	1.44	0.48	77.21	91.57	−1.01	15.05
TT60-15	2.23	0.49	72.58	93.42	−0.98	14.02
TT100-10	4.55	0.41	98.68	87.09	0.94	18.16
TT100-15	2.06	0.30	97.49	84.95	1.35	19.54

H% (d.b.) moisture content; a_w water activity at 25 °C; ME microencapsulation efficiency. Mean values (n = 2). The coefficients of variation were less than 10%.

Table 3. Parameters of the microcapsules obtained by freeze–drying.

Microcapsules	H (% d.b.)	a_w (25 °C)	ME (%)	Color		
				L*	a*	b*
STT-10	0.02	0.37×10^{-3}	99.38	97.96	−0.67	6.59
STT-15	1.33	0.41×10^{-3}	99.03	97.96	−0.57	7.86
TT60-10	0.09	0.29×10^{-3}	99.04	98.47	−0.62	5.58
TT60-15	3.00	0.47×10^{-3}	99.59	97.92	−0.40	7.31
TT100-10	0.04	0.24×10^{-3}	99.06	97.67	−0.08	7.40
TT100-15	0.03	0.33×10^{-3}	98.79	96.44	−0.49	11.32

H% (d.b.) moisture content; a_w water activity at 25 °C; ME microencapsulation efficiency. Mean values (n = 2). The coefficients of variation were less than 10%.

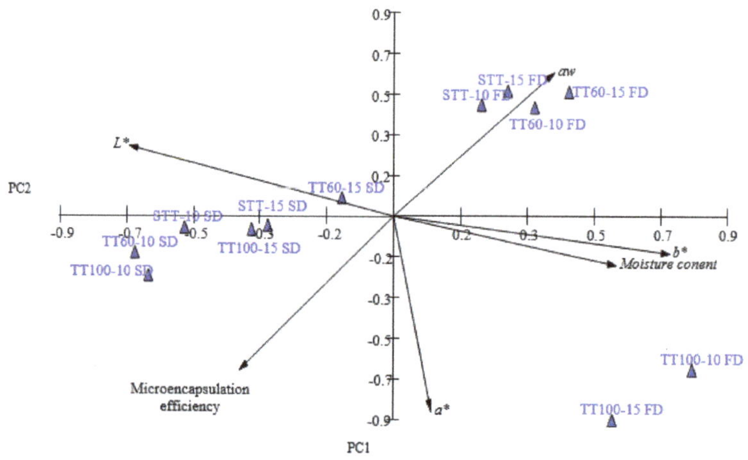

Figure 1. Principal component analysis (PCA) of chia oil microcapsules obtained from spray–drying or freeze–drying (see system codes in Table 1).

The principal components (PC) 1 and 2 explain 46.7 and 38.1% of the observed variability, respectively, totaling around 85%. As can be seen, PC1 allowed separation of the microcapsules obtained by spray-drying from those freeze-dried, while PC2 allowed discrimination within the microencapsulated chia oil by freeze-drying, the systems whose aqueous phases were treated at 100 °C, 30 min of the rest of the microcapsules. Thus, taking PC1 into account, the powders obtained by spray-drying appeared associated with the highest microencapsulation efficiencies and greatest luminosity. On the other hand, microencapsulated oils by freeze-drying were associated with higher moisture levels, water activity, and the color parameter b* (more yellowish). Regarding PC2, the microcapsules obtained by freeze-drying from parental emulsions whose aqueous phases received the most intense heat treatment were associated with a higher a* value, a higher microencapsulation efficiency, and a lower level of a_w.

4. Conclusions

The multivariate analysis made it possible to differentiate the microcapsules obtained by spray-drying from the freeze-drying ones, presenting the former ones greater luminosity and microencapsulation efficiency, and lower moisture content, aqueous activity, and b* values.

Institutional Review Board Statement: Not applicable.

Informed Consent Statement: Not applicable.

Data Availability Statement: Not applicable.

Acknowledgments: This work was supported by the grants Ia ValSe-Food-CYTED (119RT0567), Spain, Universidad Nacional de La Plata (UNLP) (11/X907), and the Agencia Nacional de Promoción Científica y Tecnológica (PICT 2016-0323), Argentina.

References

1. Eratte, D.; Dowling, K.; Barrow, C.J.; Adhikari, B. Recent advances in the microencapsulation of omega-3 oil and probiotic bacteria through complex coacervation: A review. *Trends Food Sci. Technol.* **2018**, *71*, 121–131. [CrossRef]
2. Kaushik, P.; Dowling, K.; Barrow, C.J.; Adhikari, B. Microencapsulation of omega-3 fatty acids: A review of microencapsulation and characterization methods. *J. Funct. Foods* **2015**, *19*, 868–881. [CrossRef]
3. Lane, K.E.; Derbyshire, E.J. Omega-3 fatty acids—A review of existing and innovative delivery methods. *Crit. Rev. Food Sci. Nutr.* **2017**, *58*, 62–69. [CrossRef]
4. Ixtaina, V.Y.; Martínez, M.L.; Spotorno, V.; Mateo, C.M.; Maestri, D.M.; Diehl, B.W.K.; Nolasco, S.M.; Tomás, M.C. Characterization of chia seed oils obtained by pressing and solvent extraction. *J. Food Compos. Anal.* **2011**, *24*, 166–174. [CrossRef]
5. Drusch, S.; Serfert, Y.; Scampicchio, M.; Schmidt-Hansberg, B.; Schwarz, K. Impact of Physicochemical Characteristics on the Oxidative Stability of Fish Oil Microencapsulated by Spray-Drying. *J. Agric. Food Chem.* **2007**, *55*, 11044–11051. [CrossRef]
6. Lee, Y.Y.; Tang, T.-K.; Phuah, E.-T.; Alitheen, N.B.M.; Tan, C.P.; Lai, O.-M. New functionalities of Maillard reaction products as emulsifiers and encapsulating agents, and the processing parameters: a brief review. *J. Sci. Food Agric.* **2016**, *97*, 1379–1385. [CrossRef]
7. Waller, G.; Feather, M. *The Maillard Reaction in Foods and Nutrition*; Proc. ACS Symposium Series; American Chemical Society: Las Vegas, NV, USA, 1983; ISBN 08-412-07690.
8. Baik, M.-Y.; Suhendro, E.L.; Nawar, W.W.; McClements, D.; Decker, E.; Chinachoti, P. Effects of antioxidants and humidity on the oxidative stability of microencapsulated fish oil. *J. Am. Oil Chem. Soc.* **2004**, *81*, 355–360. [CrossRef]
9. Augustin, M.; Bhail, S.; Cheng, L.; Shen, Z.; Øiseth, S.; Sanguansri, L. Use of whole buttermilk for microencapsulation of omega-3 oils. *J. Funct. Foods* **2015**, *19*, 859–867. [CrossRef]
10. Augustin, M.A.; Sanguansri, L.; Bode, O. Maillard Reaction Products as Encapsulants for Fish Oil Powders. *J. Food Sci.* **2006**, *71*, E25–E32. [CrossRef]

biology and life sciences forum

MDPI

Proceeding Paper

Preserving and Delivery Systems of Bioactives and Functional Compounds of Chia Seed (*Salvia hispanica* L.) †

Luciana M. Julio [1], Claudia N. Copado [1], Vanesa Y. Ixtaina [1,2] and Mabel C. Tomás [1,*]

1 Centro de Investigación y Desarrollo en Criotecnología de Alimentos (CIDCA), CCT-CONICET, CICPBA, Universidad Nacional de La Plata (UNLP), 47 y 116, La Plata 1900, Argentina; luci_julio86@hotmail.com (L.M.J.); claudiacopado16@gmail.com (C.N.C.); vanesaix@hotmail.com (V.Y.I.)

2 Facultad de Ciencias Agrarias y Forestales (FCAyF, UNLP), 60 y 119, La Plata 1900, Argentina

* Correspondence: mabtom@hotmail.com or mtomas@quimica.unlp.edu.ar

† Presented at the III Conference la ValSe-Food and VI Symposium Chia-Link Network, Online, 15–17 November 2021.

Abstract: There is growing interest in the development of edible delivery systems to enrich, protect and release bioactive compounds within foods. Emulsion-based systems are a good strategy for this purpose. Considering that chia oil (high levels of omega-3 fatty acids) is very susceptible to lipid oxidation, conventional and bilayer O/W emulsions were studied as a function of refrigerated storage. Monolayer emulsions were stabilized with deoiled sunflower lecithin while, in the case of bilayer ones, chitosan was also added by applying the electrostatic deposition technique. Bilayer emulsions presented a monomodal droplet size distribution while a shoulder towards larger particle sizes appeared for the conventional systems. Some signs of destabilization by the creaming process were recorded for monolayer emulsions, instead of the high stability associated with the other ones. The presence of chitosan significantly affected the rheological characteristics of emulsions by increasing their viscosity and modifying their flow behavior. In terms of oxidative stability, bilayer emulsions recorded the lowest PV values during the refrigerated storage and represent a better protective system than other ones included in the bulk oil. Thus, bilayer emulsions are a suitable option for the delivery of chia omega-3 and other PUFAs, with potential application in the food industry.

Keywords: chia by-products; electrostatic *layer-by-layer* deposition; modified sunflower lecithin; mono and bilayer O/W emulsions; omega-3 fatty acids

Citation: Julio, L.M.; Copado, C.N.; Ixtaina, V.Y.; Tomás, M.C. Preserving and Delivery Systems of Bioactives and Functional Compounds of Chia Seed (*Salvia hispanica* L.). *Biol. Life Sci. Forum* **2021**, *8*, 11. https://doi.org/10.3390/blsf2021008011

Academic Editors: Loreto Muñoz and Claudia M. Haros

Published: 25 January 2022

1. Introduction

Emulsions with enhanced stability against environmental stresses and release properties can be obtained by applying an interfacial engineering technique *layer by layer* (LBL) based on the electrostatic deposition of charged biopolymers onto droplet surfaces with opposite charge. These systems result in oil droplets being stabilized by multiple interfacial layers constituted by an emulsifier layer and one or more biopolymer ones. Different emulsion characteristics such as charge, thickness, and composition, as well as bulk physicochemical properties, can be affected by this interfacial technology. Thus, the LBL technique could be used for the food industry to develop functional food with improved stability during manufacture, storage, transport, and utilization [1]. Because of the high omega-3 content of chia oil and its susceptibility to lipid oxidation, its inclusion into multilayer systems using a natural emulsifier such as sunflower lecithin—a by-product of the oil-degumming process—and chitosan—a waste product of the marine food processing industry—is of interest.

The aim of this research work was to develop chia O/W mono- and bilayer emulsions, whilst evaluating their physicochemical characteristics as a function of refrigerated storage to obtain interesting information about their potential application as omega-3 delivery systems.

2. Materials and Methods

2.1. Materials

Chia oil was provided by Nutraceutica Sturla SRL (Buenos Aires, Argentina). Fatty acid composition was determined by GC according to IUPAC, 1992 [2], resulting in 10.5, 2.5, 6.1, 19.5 and 61.3% for C16:0, C18:0, C18:1, C18:2, C18:3, respectively. Chitosan (Ch) of medium MW∼250 kDa and deacetylation of 75–85% was purchased from Sigma Chemical Company (St. Louis, MO, USA) and sunflower lecithin was provided by Vicentin SAIC (Santa Fe, Argentina). All reagents were analytical grade.

2.2. Methods

2.2.1. Sunflower Lecithin Modification

The deoiling process of native sunflower lecithin was carried out according to American Oil Chemists' Society Official Method Ja 4–46 [3].

2.2.2. Preparation of Emulsions

A primary emulsion (L) with 5% (wt/wt) of chia seed oil and deoiled sunflower lecithin (DSL) was obtained in two homogenization steps using an Ultraturrax T-25 (Janke and Kunkel GmbH, Staufen, Germany) (9500 rpm, 1 min) and a high-pressure valve homogenizer (Niro Soavi, Parma, Italy) (600 bar, 4 passes) at pH 3. From the monolayer system, a secondary emulsion (LCh) was obtained through the addition of 0.2% chitosan. In order to prevent microbial growth both nisine 0.0012% (wt/wt) and potassium sorbate 0.1% (wt/wt) were added. Mono- and bilayer systems were stored 30 days at 4 °C.

2.2.3. Droplet Size

The droplet size distribution curves (DSD) and the $D_{[3, 2]}$ mean diameter were determined using a particle size analyzer Malvern Mastersizer 2000E (Malvern Instruments Ltd., Worcestershire, UK).

2.2.4. ζ-Potential

The ζ-potential measurements were performed with a Zeta Potential Analyzer Brookhaven 90 Plus/Bi-MAS (Brookhaven Instruments Corporation, Holtsville, NY, USA) in a range −100 to 50 mV according to Julio et al. [4].

2.2.5. Rheological Properties

Rheological measurements were carried out at 25 ± 1 °C using a rheometer Haake RS600 (Haake, Germany) with a coarse plate–plate sensor system according to Julio et al. [4].

2.2.6. Emulsion Stability

Emulsions' global stability was monitored by the evolution of their backscatter profiles using a Vertical Scan Analyzer Quick Scan (Coulter Corp., Miami, FL, USA) according to Pan et al. [5]. Samples were transferred to cylindrical glass tubes and measured periodically for 30 days.

2.2.7. Peroxide Value (PV)

The primary products of lipid oxidation were determined according to Shantha and Decker [6] as a function of refrigerated storage.

3. Results and Discussion

Figure 1 shows the schemes of the monolayer (L) and bilayer (LCh) systems with their respective ζ– potential values at pH 3. The electrostatic deposition of cationic chitosan onto the DSL-stabilized oil droplets was evidenced thought the inversion charge. In this sense, the droplet surface charge of monolayer emulsions of −37 mV turned to +46 mV for bilayer emulsions because of chitosan deposition. These results could be due to the opposite charges between DSL and chitosan molecules at this pH level, both being compounds mutually attracted.

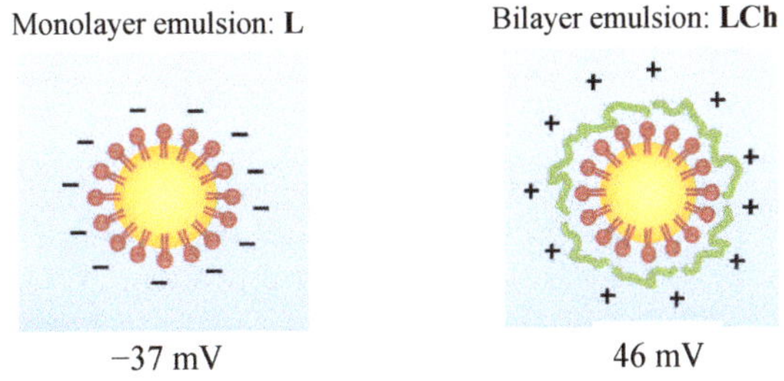

Figure 1. Schemes of chia oil droplets from mono- and bilayer emulsions and their ζ-potential values.

Bilayer emulsions presented a monomodal DSD curve while a shoulder corresponding to a second population of larger droplet sizes was observed in DSD from monolayer systems (data not shown); $D_{[2, 3]}$ values for L and LCh emulsions were 0.32 and 0.26 μm, respectively. No significant ($p > 0.05$) changes in particle size were recorded during refrigerated storage.

The flow curves of the different systems were obtained from the rheological data fitting to the Power Law model equation ($R^2 > 0.99$). The apparent viscosity values of emulsions at 100 s − 1 (η100)—typical shear rate for food processes, such as stirring or mastication, flow through a pipe—were also calculated [7]. According to Figure 2, monolayer systems behaved as Newtonian fluids (n~1), while the chitosan deposition into bilayer emulsions led to shear-thinning behavior ($n < 1$). Additionally, a significant increase ($p < 0.05$) in η_{100} was recorded, resulting in 1.54 and 9.97×10^{-3} Pa.s^n for L and LQ systems, respectively.

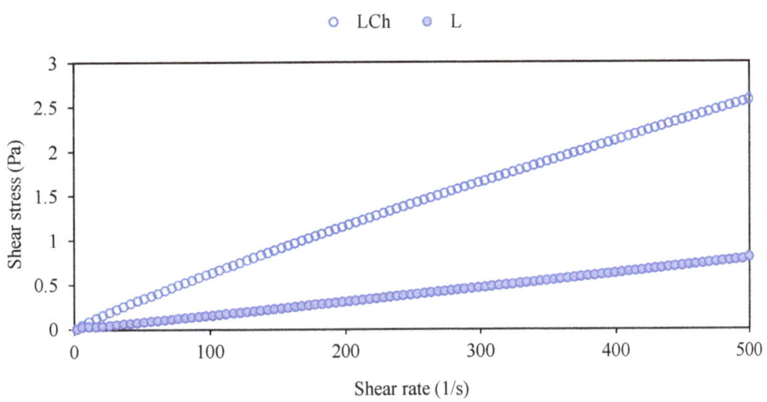

Figure 2. Flow curves of mono- and bilayer emulsions with chia oil.

Regarding global stability, primary emulsions experienced creaming destabilization, which can be seen from their BS profiles. In contrast, secondary systems were more stable, with their BS profiles remaining unchanged over the storage period. This fact could be associated with the higher aqueous phase viscosity of bilayer emulsions, which would reduce the oil droplets' movement (Figure 3).

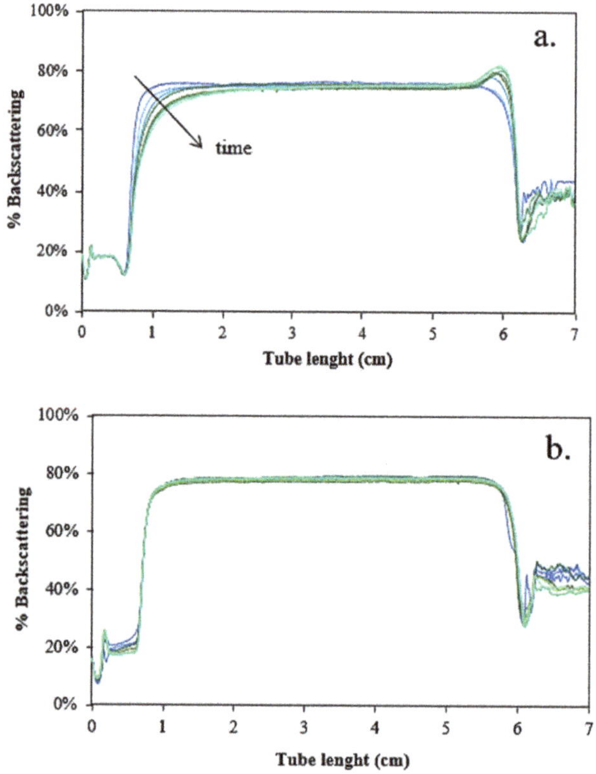

Figure 3. Back scattering profiles (%BS vs. tube length) of mono- (**a**) and bilayer (**b**) chia O/W emulsions during refrigerated storage.

Bilayer emulsions with DSL and chitosan presented a lower PV and, therefore, higher oxidative stability than bulk oil and monolayer systems after 30 days of refrigerated storage (Figure 4). This fact indicates the positive effect of the bilayer formed around the oil droplets to improve the protection against lipid oxidation. Some authors suggest that chitosan could act as a free radical scavenger, while others indicate that positively charged droplets could repel Fe^{+} ions, reducing lipid oxidation [8].

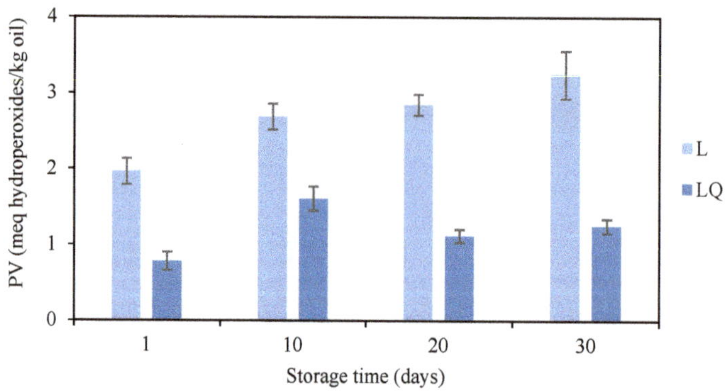

Figure 4. Evolution of hydroperoxides values of mono- and bilayer chia O/W emulsions at 1, 10, 20 and 30 days of refrigerated storage. Average values are shown (n = 2).

4. Conclusions

Mono- and bilayer emulsions containing chia oil were developed by applying the *layer-by-layer* technique. The electrostatic deposition of chitosan onto a DSL interfacial membrane to form a bilayer around the chia oil droplets was evidenced through the inversion charge at pH 3.

Bilayer emulsions presented smaller and monodisperse oil droplets and a better long-term stability than conventional systems. Additionally, bilayer emulsions provided higher protection against chia oil oxidation, recording lower PV values over the storage period. From these results, the bilayer emulsion proved to be an effective system to protect and deliver of omega-3 fatty acids from chia oil into functional foods.

Funding: This research received no external funding.

Institutional Review Board Statement: Not applicable.

Informed Consent Statement: Not applicable.

Data Availability Statement: Not applicable.

Acknowledgments: This work was supported by grant Ia ValSe-Food-CYTED (119RT0567) Spain, Universidad Nacional de La Plata (UNLP) (11/X907), and the ANPCyT (PICT 2016-0323), Argentina.

Conflicts of Interest: The authors declare no conflict of interest.

References

1. Klinkesorn, U.; Sophanodora, P.; Chinachoti, P.; McClements, D.J.; Decker, E.A. Stability of Spray-Dried Tuna Oil Emulsions Encapsulated with Two-Layered Interfacial Membranes. *J. Agric. Food Chem.* **2005**, *53*, 8365–8371. [CrossRef] [PubMed]
2. IUPAC. Standard methods for the analysis of oils, fats and derivates. In *International Union of Pure and Applied Chemistry, Oxford*, 7th ed.; Paquot, C., Hautfenne, A., Eds.; Blackwell Scientific Publications Inc.: Oxford, UK, 1992.
3. AOCS. *Official Methods and Recommended Practices of the American Oil Chemists' Society*, 4th ed.; AOCS Press: Champaign. IL, USA, 1994.
4. Julio, L.M.; Copado, C.N.; Diehl, B.W.; Ixtaina, V.Y.; Tomás, M.C. Chia bilayer emulsions with modified sunflower lecithins and chitosan as delivery systems of omega-3 fatty acids. *LWT* **2018**, *89*, 581–590. [CrossRef]
5. Pan, L.G.; Tomás, M.; Añon, M.C. Effect of sunflower lecithins on the stability of water-in-oil and oil-in-water emulsions. *J. Surfactants Deterg.* **2002**, *5*, 135–143. [CrossRef]
6. Shantha, N.C.; Decker, E. Rapid, Sensitive, Iron-Based Spectrophotometric Methods for Determination of Peroxide Values of Food Lipids. *J. AOAC Int.* **1994**, *77*, 421–424. [CrossRef] [PubMed]
7. McClements, D.J. *Food Emulsions: Principles, Practices, and Techniques*; CRC Press: Boca Raton, FL, USA, 2004.
8. Klinkesorn, U.; Sophanodora, P.; Chinachoti, P.; McClements, D.J.; Decker, E.A. Increasing the Oxidative Stability of Liquid and Dried Tuna Oil-in-Water Emulsions with Electrostatic Layer-by-Layer Deposition Technology. *J. Agric. Food Chem.* **2005**, *53*, 4561–4566. [CrossRef] [PubMed]

biology and life sciences forum

Proceeding Paper

Evaluation of Functional and Nutritional Properties of Hydrolyzed Broad Bean and Quinoa Flours †

Ileana de los A. Gremasqui, Maria A. Giménez, Manuel O. Lobo and Norma C. Sammán *

Centro de Investigación Interdisciplinario en Tecnología y Desarrollo Social del NOA (CIITED), Consejo Nacional de Investigaciones Científicas y Técnicas (CONICET), Facultad de Ingeniería, Universidad Nacional de Jujuy, Italo Palanca 10, Jujuy CP 4600, Argentina; ileanagremasqui96@gmail.com (I.d.l.A.G.); malejandragimenez@gmail.com (M.A.G.); mlobo958@gmail.com (M.O.L.)

* Correspondence: normasamman@gmail.com

† Presented at the III Conference la ValSe-Food and VI Symposium Chia-Link Network, Online, 15–17 November 2021.

Abstract: In sports nutrition, protein intake is essential to stimulate protein synthesis and repair muscle damage caused by exercise. The search for non-traditional protein sources has increased in recent years. Quinoa *(Chenopodium quinoa* Wild*)* and broad beans *(Vicia faba* L.*)* grains could be used in the production of protein products. Broad beans are an introduced and widely expanded crop in South America; it is part of the Argentine Northwest Andean population diet. The aim of this work was to evaluate the functional and nutritional properties of hydrolyzed quinoa (HQF) and broad bean (HBF) flours for their use in the elaboration of protein foods for athletes. Both hydrolyzed flours were obtained using Flavourzyme at 50 °C and pH 8 for 3 and 1 h, respectively. HQF presented a higher degree of hydrolysis (21.79%), while HBF had higher protein content (57.31%), yield (32.14%), and protein recovery (71.31%). In HBF and HQF, Na and K were the most abundant minerals, both necessary for the replacement of electrolytes lost during physical training. HBF and HQF presented 5909.63 and 2708.91 mg/100 g of properties, respectively, and HQF presented higher emulsifying branched amino acids content, essential in sports nutrition. Regarding technological activity (61.30 m^2/g), stability indexes (158.6 min), and foaming capacity (131%); HBF shows a wider range of solubility in function of pH, and good foaming stability (68–92%). These results indicate that HQF and HBF could be potential ingredients for athletes' protein supplements formulation.

Keywords: broad beans; flour; functional properties; hydrolysis; nutritional properties; quinoa

Citation: Gremasqui, I.d.l.A.; Giménez, M.A.; Lobo, M.O.; Sammán, N.C. Evaluation of Functional and Nutritional Properties of Hydrolyzed Broad Bean and Quinoa Flours. *Biol. Life Sci. Forum* **2021**, *8*, 12. https://doi.org/10.3390/blsf2021008012

Academic Editors: Loreto Muñoz and Claudia Monica Haros

Published: 18 February 2022

1. Introduction

In sports nutrition, the consumption of hydrolyzed protein supplements has gained relevance due to their ability to modulate the anabolism of skeletal muscle proteins, improve sports performance, influence the control of body mass, and accelerate the digestion and absorption of proteins, increasing the availability of amino acids [1]. These products are obtained from protein concentrates or isolates by enzymatic hydrolysis. Obtaining hydrolysates from vegetable flours could provide advantages in the functional and nutritional properties of the product obtained due to the presence of fibers and minerals in the flours [2]. The organoleptic quality can also be improved, due to significant changes in taste. The food industry is in search of alternative protein sources. Broad beans (*Vicia faba* L.) are an introduced and widely expanded crop in South America; it is part of the Argentine Northwest Andean population diet and quinoa (*Chenopodium quinoa* Wild) is one of the most nutritive Andean grains. Due to their properties, quinoa (*Chenopodium quinoa* Wild) and broad beans (*Vicia faba* L.) grains could be used in the production of these protein products. The aim of this work was to evaluate the functional and nutritional properties of hydrolyzed quinoa (HQF) and broad bean (HBF) flours for their use in the elaboration of protein foods for athletes.

2. Materials and Methods

2.1. Materials

Broad bean (BF) (*Vicia faba* L.) and quinoa (QF) (*Chenopodium quinoa* Wild) flours were provided by producers from Quebrada de Humahuaca, Jujuy. Quinoa flour was previously defatted.

2.2. Obtaining Hydrolyzed Flours and Degree of Hydrolysis (DH)

The hydrolyzed flours were prepared according to Lee et al. [3]. The enzyme used was Flavourzyme (Sigma Aldrich, St. Louis, MO, USA, 25 LAPU/g proteins) at 50 °C, pH 8, 1 and 3 h for BF and QF, respectively. The mixtures obtained were centrifuged at 4500× g/30 min. The enzymes were inactivated at 85 °C and then the pH was adjusted to 7. The degree of hydrolysis was determined with trichloroacetic acid (TCA). The supernatants were dried at low temperature and ground to a particle size <149 µm.

2.3. Performance Parameters and Chemical and Nutritional Properties

Mass yield and protein recovery were determined according to Noman et al. [4]. AOAC (2017) methods were used to determine the content of proteins, lipids, ash, soluble dietary fiber, and minerals. For soluble sugars, the method of Dubois et al. [5] method was used, and for amino acids, the method of Mota et al. [6].

2.4. Functional Properties of Hydrolyzed Flours

Protein solubility, emulsifying activity index (EAI), emulsion stability index (ESI), foaming capacity (FC) and foaming stability (FS) were determined in hydrolyzed flour according to Gremasqui et al. [7].

2.5. Statistical Analysis

The INFOSTAT program was used for the analysis of variance of the data and Fisher's LSD to compare the means with a significance level of 0.05.

3. Results and Discussion

3.1. Degree of Hydrolysis (DH)

The DH values of HBF and HQF are shown in Table 1. The lower DH obtained in HBF could be due to the shorter hydrolysis time used and the release of larger peptides insoluble in TCA. According to Barac et al. [8], hydrolysates with DH > 10% are characterized by good nutritive values.

Table 1. Performance parameters and nutritional composition of HBF and HQF.

Parameters #	HBF	HQF
DH	12.57 ± 0.96^{a}	21.79 ± 0.78^{b}
Mass yield	32.80 ± 0.93^{b}	21.03 ± 1.22^{a}
Protein recovery	71.32 ± 2.03^{b}	67.28 ± 1.73^{a}
Protein content	57.31 ± 0.27^{b}	54.69 ± 0.12^{a}
Lipids content	0.004 ± 0.000^{a}	0.003 ± 0.000^{a}
Ash content	11.79 ± 0.24^{a}	14.11 ± 0.28^{b}
SDF	10.56 ± 0.41^{a}	23.18 ± 0.74^{b}
HC dig.	20.34	8.02
Soluble sugars	9.95 ± 0.87^{a}	9.35 ± 0.89^{a}
Na *	2376.35 ± 39.60^{a}	3121.96 ± 11.65^{b}
Cu *	2.18 ± 1.04^{a}	2.18 ± 0.18^{a}
Fe *	18.04 ± 3.44^{a}	20.25 ± 0.44^{b}
Zn *	9.26 ± 4.08^{b}	7.45 ± 0.57^{a}
Ca *	411.79 ± 5.66^{b}	130.46 ± 1.60^{a}

Table 1. *Cont.*

Parameters #	HBF	HQF
Mg *	344.27 ± 1.99 b	180.27 ± 7.58 a
K *	3526.42 ± 195.20 a	4000.27 ± 135.70 b

The values correspond to means ± deviations and are expressed in (g/100 g db), n = 3; * mg/100 g db. Different letters in each row are significantly different ($p < 0.05$).

3.2. Performance Parameters and Nutritional Composition

The mass yield, protein recovery, and chemical and nutritional composition of hydrolyzed flours are shown in Table 1. HBF had higher mass yield and protein recovery than HQF ($p < 0.05$) due to their higher natural bean protein content; this agrees with the results found by Thamnarathip et al. [9]. The high protein content of hydrolyzed flours (>50%) makes them useful to be incorporated into sports foods. Soluble dietary fiber (SDF) content was significantly different ($p < 0.05$) between both hydrolyzed flours.

Na and K were the most abundant minerals in both hydrolyzed flours, both necessary for the replacement of electrolytes lost during physical training [10]. According to Xia et al. [10], minerals such as Fe, Zn, and Mg also stand out for their usefulness for an athlete's nutrition. A portion of 30 g of these hydrolyzed flours would contribute between 73% to 76% of the recommended protein dose (20–25 g) to stimulate muscle protein synthesis after exercise [1]; 12–25% of suggested fiber dietary intake (25–30 g/day); approximately 48–62% and 53–60% of Na and K RDI, respectively, for athletes.

Table 2 shows the total and free amino acid composition of hydrolyzed flours. The total essential amino acid content was 6415.13 and 11901.71 mg/100 g of HQF and HBF, respectively. The predominant essential amino acids were leucine, lysine, and valine; these results agree with Muhamyankaka et al. [2]. HBF and HQF presented high contents of branched amino acids, essential in sports nutrition to improve sports performance by reducing the appearance of fatigue. Both hydrolyzed flours are rich in glutamic acid, glutamine precursor, which is important to recover muscle glycogen deposits and avoid loss of muscle mass.

Table 2. Amino acid composition of HBF and HQF.

Amino Acid (mg/100 g Sample)	HBF a	HQF a	HBF b	HQF b
Essential amino acid				
Histidine	855.3 ± 25.0 b	639.3 ± 8.2 a	26.54 ± 1.53 a	116.63 ± 2.12 b
Threonine	1375.5 ± 35.9 b	760.2 ± 3.9 a	13.54 ± 4.35 a	27.56 ± 2.95 b
Valine	1578.6 ± 20.4 b	837.5 ± 1.1 a	20.04 ± 4.57 a	41.69 ± 4.77 b
Methionine	292.1 ± 12.5 a	434.7 ± 5.3 b	24.91 ± 3.32 a	50.46 ± 4.02 b
Lysine	1928.8 ± 6.7 b	1024.6 ± 10.3 a	n.d	n.d
Isoleucine	1344.9 ± 35.5 b	642.1 ± 0.2 a	15.95 ± 0.44 a	35.78 ± 7.07 b
Leucine	2986.1 ± 78.8 b	1229.3 ± 1.3 a	27.19 ± 4.06 a	99.48 ± 4.84 b
Phenylalanine	1540.4 ± 63.0 b	847.4 ± 1.9 a	49.48 ± 10.81 a	149.82 ± 2.60 b
Tryptophan	n.d	n.d	33.31 ± 2.12 a	91.36 ± 5.13 b
Nonessential amino acid				
Aspartic acid	6030.4 ± 63.4 b	2384.3 ± 33.1 a	206.48 ± 2.17 b	17.23 ± 2.83 a
Glutamic acid	9475.7 ± 113.9 b	5475.6 ± 36.2 a	69.26 ± 4.72 a	87.68 ± 7.30 b
Serine	2129.2 ± 57.0 b	1023.0 ± 4.8 a	11.83 ± 1.21 a	17.47 ± 2.83 b
Arginine	3409.2 ± 115.0 b	1624.3 ± 15.5 a	429.52 ± 7.07 b	41.50 ± 0.70 a
Cysteine	173.2 ± 21.3 a	179.4 ± 8.3 b	n.d	n.d
Tyrosine	1338.0 ± 49.5 b	772.5 ± 9.7 a	40.77 ± 0.89 a	95.75 ± 1.55 b
Glycine	1699.5 ± 58.4 b	1275.5 ± 23.5 a	22.65 ± 3.62 a	35.80 ± 0.69 b
Alanine	1921.6 ± 23.4 b	1139.5 ± 7.5 a	46.81 ± 1.12 a	118.95 ± 3.07 b

Table 2. *Cont.*

Amino Acid (mg/100 g Sample)	HBF [a]	HQF [a]	HBF [b]	HQF [b]
Proline	1917.7 ± 39.7 [b]	919.2 ± 6.9 [a]	66.29 ± 4.01 [a]	75.31 ± 3.06 [b]
Total amino acid	39996.2 ± 794.5 [b]	21208.5 ± 85.5 [a]	1104.56 ± 17.46 [b]	1102.48 ± 8.98 [a]
Taste component [c]				
Umami			275.74 ± 6.89 [b]	104.92 ± 10.13 [a]
Sweet			501.69 ± 2.63 [b]	205.48 ± 2.25 [a]
Bitter			253.05 ± 3.21 [a]	604.97 ± 3.29 [b]

[a] Total amino acid content, [b] Free amino acid content, [c] Umami taste: Glu + Asp; Sweet taste: Ala + Arg + Ser + Thr; Bitter taste: Gly + His + Ile + Leu + Met + Phe + Val + Pro; n.d: not detected.

HBF was rich in free arginine (38.88%) and aspartic acid (18.69%), while HQF was rich in free phenylalanine (13.60%), alanine (10.80%), histidine (10.57%), and leucine (9.02%); these results are similar to those found by Laohakunjit et al. [11]. The free amino acids would significantly affect the taste characteristics of hydrolyzed flours. HBF had the highest amount of umami and sweet amino acids, whereas HQF had the highest content of bitter amino acids.

3.3. Functional Properties

Protein solubility. Figure 1 shows HBF and HQF protein solubility. HBF showed significantly higher solubility and a greater range of variation with the pH (65.10–87.54%) than HQF ($p < 0.05$). The minimum protein solubility was at pH 4 (16.22 and 33.77% for HBF and HQF, respectively). The higher protein solubility in HBF could be due to its greater exposure of polar amino acid that interacts with water through hydrogen bonding. The inconsistency between the lower DH and higher solubility in HBF could be due to the balance between hydrophilic and hydrophobic forces scores over DH.

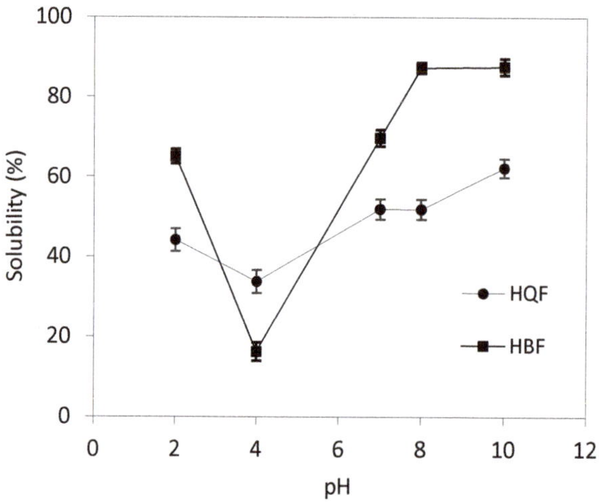

Figure 1. Protein solubility of HBF (■) and HQF (•) as a function of pH.

Emulsifying properties. HQF presented better surfactant properties than HBF ($p < 0.05$) (Table 3). The lower EAI and ESI values in HBF could be due to the lower DH and a higher exposure of hydrophilic groups that would bind with peptides in the aqueous phase, decreasing hydrophobicity and emulsifying stability [12].

Table 3. Functional properties of HBF and HQF.

Functional Properties	HBF [#]	HQF [#]
EAI (m^2/g)	38.57 ± 1.45 [a]	61.30 ± 0.14 [b]
ESI (min)	55.01 ± 3.56 [a]	158.60 ± 1.70 [b]
FC (%)	89.50 ± 6.36 [a]	131.00 ± 4.24 [b]

[#] The values correspond to means ± standard deviations, n = 3. Different letters in each row indicate significant differences ($p < 0.05$).

Foaming properties. According to the results shown in Table 3, HQF has the highest CF value, probably due to the higher DH responsible for a greater number of small peptides that are easily adsorbed at the air-water interface [2]. Figure 2 shows the stability of the foam (FS). HBF and HQF showed significant differences ($p < 0.05$) after 10 min, HBF showing better FS [12].

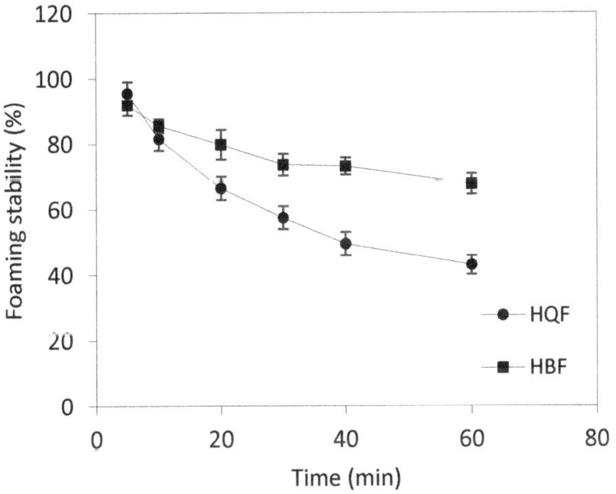

Figure 2. Foaming stability of HBF (■) and HQF (•) as a function of time.

4. Conclusions

The enzymatic hydrolysis of quinoa and broad bean flours has been an adequate way to improve the nutritional properties (high protein content and good source of essential and branched amino acids), making them suitable as ingredients in the preparation of food for athletes. HBF and HQF were characterized by having higher levels of free amino acids that produce sweet and sour tastes. On the other hand, the hydrolyzed products presented high solubility and good surfactant properties, which is why they could be used in the preparation of beverages, creams, butter, ice cream, mousses, and cakes. The enzymatic hydrolysis applied directly to the flours caused a positive effect on the nutritional and functional properties.

Author Contributions: Conceptualization, and methodology, I.d.l.A.G. and M.A.G.; validation I.d.l.A.G., M.A.G., M.O.L. and N.C.S.; formal analysis, I.d.l.A.G. and M.A.G.; investigation, I.d.l.A.G. and M.A.G.; resources, M.O.L. and N.C.S.; data curation, I.d.l.A.G.; writing—original draft preparation, I.d.l.A.G.; writing—review and editing, M.A.G., M.O.L. and N.C.S.; visualization, I.d.l.A.G.; supervision, M.A.G., M.O.L. and N.C.S.; project administration, M.O.L. and N.C.S.; funding acquisition, M.O.L. and N.C.S. All authors have read and agreed to the published version of the manuscript.

Funding: This research received no externa funding.

Institutional Review Board Statement: Not applicable.

Informed Consent Statement: Not applicable.

Data Availability Statement: Not applicable.

Acknowledgments: This work was supported by grant Ia ValSe-Food-CYTED (Projet N° 119RT0567) and SECTER—Universidad Nacional de Jujuy—CONICET.

Conflicts of Interest: The authors declare no conflict of interest.

References

1. de Antuñano, N.P.G.; Marqueta, P.M.; Redondo, R.B.; Fernández, C.J.C.; de Teresa Galván, C.; del Valle Soto, M.; Bonafonte, L.F.; Gabarra, A.G.; Gaztañaga, T.A.; Manonelles, P.M.; et al. Suplementos nutricionales para el deportista. Ayudas ergogénicas en el deporte-2019. Documento de consenso de la Sociedad Española de Medicina del Deporte. *Arch. Med. Deporte* **2019**, *36*, 1–114.
2. Muhamyankaka, V.; Shoemaker, C.F.; Nalwoga, M.; Zhang, X.M. Physicochemical properties of hydrolysates from enzy-matic hydrolysis of pumpkin (Cucurbita moschata) protein meal. *Int. Food Res. J.* **2013**, *20*, 2227–2240.
3. Lee, J.-Y.; Lee, H.D.; Lee, C.-H. Characterization of hydrolysates produced by mild-acid treatment and enzymatic hydrolysis of defatted soybean flour. *Food Res. Int.* **2001**, *34*, 217–222. [CrossRef]
4. Noman, A.; Qixing, J.; Xu, Y.; Ali, A.; Al-Bukhaiti, W.Q.; Abed, S.M.; Xia, W. Influence of Degree of Hydrolysis on Chemical Composition, Functional Properties, and Antioxidant Activities of Chinese Sturgeon (Acipenser sinensis) Hydrolysates Obtained by Using Alcalase 2.4 L. *J. Aquat. Food Prod. Technol.* **2019**, *28*, 583–597. [CrossRef]
5. Dubois, M.; Gilles, K.A.; Hamilton, J.K.; Rebers, P.A.; Smith, F. Colorimetric Method for Determination of Sugars and Related Substances. *Anal. Chem.* **1956**, *28*, 350–356. [CrossRef]
6. Mota, C.; Santos, M.; Mauro, R.; Samman, N.; Matos, A.S.; Torres, D.; Castanheira, I. Protein content and amino acids profile of pseudocereals. *Food Chem.* **2014**, *193*, 55–61. [CrossRef] [PubMed]
7. Gremasqui, I.A.; Giménez, M.A.; Lobo, M.O.; Sammán, N.C. Nutritional and Functional characterisation of hydrolysates from quinoa flour (*Chenopodium quinoa*) using two proteases. *Int. J. Food Sci. Technol.* **2021**, *56*, 6507–6514. [CrossRef]
8. Barać, M.; Čabrilo, S.; Pešić, M.; Stanojević, S.; Pavlićević, M.; Maćej, O.; Ristić, N. Functional Properties of Pea (Pisum sativum, L.) Protein Isolates Modified with Chymosin. *Int. J. Mol. Sci.* **2011**, *12*, 8372–8387. [CrossRef]
9. Thamnarathip, P.; Jangchud, K.; Jangchud, A.; Nitisinprasert, S.; Tadakittisarn, S.; Vardhanabhuti, B. Extraction and characterisation of Riceberry bran protein hydrolysate using enzymatic hydrolysis. *Int. J. Food Sci. Technol.* **2015**, *51*, 194–202. [CrossRef]
10. Xia, R. Study on the Reasonable Supplement of Vitamin and Minerals for Athletes. *Adv. J. Food Sci. Technol.* **2015**, *8*, 303–305. [CrossRef]
11. Laohakunjit, N.; Kerdchoechuen, O.; Kaprasob, R.; Matta, F.B. Volatile Flavor, Antioxidant Activity and Physicochemical Properties of Enzymatic Defatted Sesame Hydrolysate. *J. Food Process. Preserv.* **2016**, *41*, e13075. [CrossRef]
12. Betancur-Ancona, D.; Martínez-Rosado, R.; Corona-Cruz, A.; Castellanos-Ruelas, A.; Jaramillo-Flores, M.E.; Chel-Guerrero, L. Functional properties of hydrolysates fromPhaseolus lunatusseeds. *Int. J. Food Sci. Technol.* **2008**, *44*, 128–137. [CrossRef]

biology and life sciences forum MDPI

Proceeding Paper

Chenopodium quinoa to Modulate Innate Myeloid Cells in the Induction of Obesity †

José Moisés Laparra [1,*], Elena Aguilar-Aguilar [2] and Claudia Monika Haros [3]

1 Molecular Immunonutrition Group, Madrid Institute for Advanced Studies in Food (IMDEA-Food), Ctra. de Canto Blanco No 8, 28049 Madrid, Spain
2 Nutrition and Clinical Trials Unit, GENYAL Platform IMDEA-Food Institute, CEI UAM + CSIC, 28049 Madrid, Spain; elena.aguilar@imdea.org
3 Instituto de Agroquímica y Tecnología de Alimentos (IATA), Consejo Superior de Investigaciones Científicas (CSIC), Av. Agustín Escardino 7, Parque Científico, Paterna, 46980 Valencia, Spain; cmharos@iata.csic.es
* Correspondence: moises.laparra@imdea.org

† Presented at the III Conference la ValSe-Food and VI Symposium Chia-Link Network, online, 15–17 November 2021.

Abstract: Complex interactions between innate and adaptive immune effectors are an important component in the induction of obesity. Particularly, different subsets of myeloid cells play key roles in metabolic liver diseases and, therefore, are promising targets for intervention strategies. *Chenopodium quinoa* seeds constitute a good source of immunonutritional compounds, which help prevent high-fat, diet-enhanced innate immune signaling via TLR4/MyD88 that boosts inflammation. Herein, two metabolic mouse models—wild type (WT) and tributyltin treated (TBT)—were used to examine the effects associated with non-alcoholic fatty liver disease (NAFLD); mice were fed with a high-fat diet (HFD) and administered with wheat or *C. quinoa* bread. Variations in myeloid cells were obtained from a hemogram analysis, and rt-qPCR (mRNA) served to evaluate macrophage markers (i.e., CD68/CD206 ratio) as well as liver inflammation (i.e., Lyve-1) to gain insights into their selective functional differentiation into metabolically injured livers. Only administration of *C. quinoa* bread prevented alterations in the liver/body weight ratio either in WT animals or those treated with TBT. These effects were associated with significantly increased variations in the peripheral myeloid cell population. Hepatic mRNA markers revealed that *C. quinoa* enables a selective functional differentiation and function of intrahepatic monocyte-derived macrophages preserving tissue integrity and function.

Keywords: *Chenopodium quinoa*; innate myeloid cells; immunonutrition; obesity

Citation: Laparra, J.M.; Aguilar-Aguilar, E.; Haros, C.M. *Chenopodium quinoa* to Modulate Innate Myeloid Cells in the Induction of Obesity. *Biol. Life Sci. Forum* **2021**, *8*, 13. https://doi.org/10.3390/blsf2021008013

Academic Editor: Loreto Muñoz

Published: 14 January 2022

1. Introduction

Obesity is recognized as overweight caused by the dysfunctional accumulation of energy reserves as fat depots, and its prevalence appears associated with an increased incidence of metabolic disorders. Complex interactions between innate and adaptive immune effectors are an important component in the induction of obesity. Accordingly, recent research demonstrated that myeloid cells accumulate in the liver as monocytes and macrophages during the progression of obesity-related non-alcoholic fatty liver disease to steatohepatitis [1]. These cells contribute to either worsening or improving tissue homeostasis following impairment of liver function. Specific environmental signals within the gut–liver axis further determine the selective functional differentiation and function of hepatic macrophages. Different subsets of these myeloid cells have pivotal roles in metabolic liver diseases; thus, they provide promising targets for intervention strategies with a preventive and/or therapeutic application.

Under a high-fat diet, innate immune Toll-like receptor (TLR)-4/MyD88 signaling leads to an inhibited macrophage proliferation to infiltrate into adipose tissue boosting inflammation [2]. *C. quinoa* seeds constitute a good source of immunonutritional serine-type protease inhibitors (SETIs), which enable innate immune events mediated by TLR4 downstream signaling that can be associated with a delayed wave, implying adaptor molecules such as TRAM/TRIF [3,4].

In view of the pivotal role of the hepatic immune–metabolic crosstalk, thereby influencing the natural history of obesity, this study evaluates the impact of the inclusion of *C. quinoa* flour into bread formulations in the variations and polarization of the myeloid population in metabolic mouse models.

2. Materials and Methods

2.1. Metabolic Mouse Models

C57BL/6 mice (6 weeks of age) born from untreated females or receiving obesogen tributyltin (50 nM) via drinking water to develop a state of NAFLD were used [5]. Afterward, both F1_A and F1_B generations were kept under an HFD until reaching 6 weeks of age. Bread formulations were administered (14 mg/day/animal) to the different animal models 3 times per week for 3 weeks.

Animal experiments were carried out in strict accordance with the recommendations in the Guide for the Care and Use of Laboratory Animals of CSIC (Consejo Superior de Investigaciones Científicas), and the protocol was approved by its ethics committee (Proex No. 080/19).

2.2. Hemogram

The whole blood count was performed on an automated hemocytometer (Abacus Junior Vet, ELECTROMEDINTER SL).

2.3. Markers of Selective Functional Differentiation Intrahepatic Macrophages

Validated gene for murine CD68 (M1-like phenotype) (forward 5′- AGA AGT GCA ATG GTG GGT CT-3′, reverse 5′- TGG GGC TTA AAG AGG GCA AG -3′), CD206 (M2-like phenotype) (forward 5′- TGC AAG CTT GTA GGA AGG AGG -3′, reverse 5′- GAT TAG AGT GGT GAG CAG GC -3′); Lyve-1 (forward 5′- CCC TCC ATT ACC AGT TGT CCC -3′, reverse 5′- ACG GCT CAT CAT CAC CAT TCT C -3′), and β-actin (forward 5′- GGC TCC TAG CAC CAT GAA GAT CAA -3′, reverse 5′- AGC TCA GTA ACA GTC CGC CTA GAA -3′) was purchased from Applied Biosystems (Foster City, CA, USA). RT-qPCR was performed with 500 ng of cDNA from liver sections, using the Universal PCR Master Mix (Applied Biosystems, ThermoFisher®). Quantitative values were calculated by using the $2^{-\Delta Ct}$ method [3].

2.4. Statistical Analyses

The statistical analysis between the different groups of treatment within the same experimental model was conducted using one-way analysis of variance (ANOVA) and the Kruskal–Wallis post hoc test by ranks. Analyses were performed with the software Statgraphics Centurion XVI, and significance was established at $p < 0.05$ for all comparisons.

3. Results

3.1. Food Intake and Morphometric Measurements

Animals administered with either wheat or *C. quinoa* bread formulations displayed reduced consumption rates in relation to controls (Figure 1A. With *C. quinoa* bread, the food intake rate did not reach statistical significance between animals under different treatments (i.e., WT vs. TBT). Upward trends for food intake rates in TBT-affected animals could reflect decreased nutrient utilization derived from NAFLD-associated liver dysfunction.

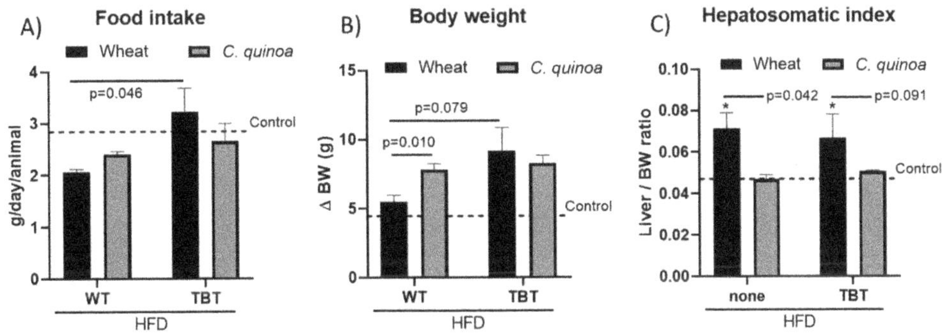

Figure 1. Food intake (**A**) and morphometric measurements, body weight gain (BW) (**B**), and hepatosomatic index (**C**) of in wild-type (WT) and tributyltin (TBT)-exposed mice fed a high-fat diet. Results are expressed as mean ± standard error (SEM) (n = 6). Untreated controls are represented by the dotted line.

Administration of *C. quinoa* bread to wild-type (WT) animals favored higher body weight gain than those fed with wheat bread (Figure 1B). However, these differences were abolished in TBT-treated animals. Notably, animals receiving *C. quinoa* bread showed similar effects in both metabolic mouse models. Administration of *C. quinoa* bread prevented alterations in the hepatosomatic index either in wild-type animals or those displaying a transgenerational inheritance to develop NAFLD-associated obesity (Figure 1C). Altogether, data may interpret the results as a differential engagement of metabolic processes mainly derived from the different compositions of immunonutritional bioactive proteins.

3.2. Variations on Innate Immune Myeloid Cell Population

Animals administered with *C. quinoa* bread displayed significantly increased variations in the myeloid cell population (Figure 2A). Only a downward trend was calculated in the hepatic infiltrating monocyte-derived macrophages in relation to that in wheat bread-fed mice (Figure 2B). In both cases, a favorable CD68/CD206 ratio was found, reflecting the more prevalent M1-like phenotype of the infiltrated cells.

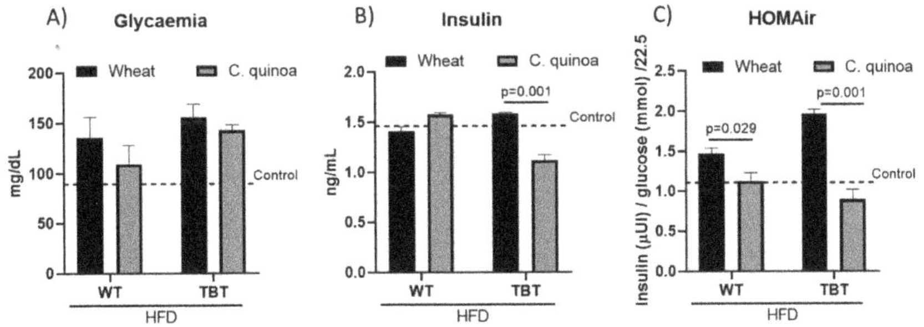

Figure 2. Innate immune adaptation: (**A**) variations in myeloid cells in wild-type (WT) and tributyltin (TBT)-exposed mice; (**B**) hepatic rt-qPCR analysis (mRNA) of macrophage marker genes CD68/CD206 and (**C**) the homing cell adhesion molecule (Lyve-1/CD44) in wild-type (WT) and tributyltin (TBT)-exposed mice fed a high-fat diet. Results are expressed as mean ± mean standard error (n = 6).

Notably, hepatic transcripts for Lyve-1 suggest the existence of two distinct functional macrophage populations when feeding *C. quinoa* or wheat bread (Figure 2C).

4. Discussion

This study investigated the associations between administration of *C. quinoa* (20%, w/w) bread, in comparison with wheat bread, and the variability of selective functional phenotypes of hepatic infiltrating monocyte-derived macrophages in metabolic mouse models with diet-induced obesity. The associations were independent of body weight gain and the CD68/CD206 proportion. The risk of cardiovascular disease and liver fibrosis in animals displaying a transgenerational inheritance to develop NAFLD was lower in those fed with *C. quinoa* bread.

Based on these findings, administration of *C. quinoa* bread enabled a better preserved hepatosomatic index, decreasing the risk of liver dysfunction and NAFLD development, which are important factors favoring the metabolic syndrome, in both metabolic mouse models [6]. The mechanism behind the associations of administration of *C. quinoa* bread and alleviation of liver inflammation and NAFLD are diverse. First, hyaluronan accumulation with HFD feeding [7], to contribute to insulin resistance and exacerbate liver inflammation by interacting with Lyve-1, is reduced by the administration of *C. quinoa* bread in WT animals. Increased insulin resistance plays a crucial role in the progression of NAFLD, which is related to adverse health outcomes. Second, preservation of Lyve-1 in TBT-treated mice, which exhibit transgenerational inheritance of disturbances in glucose homeostasis [5] and inhibition of the insulin receptor expression [8] as well as a permanently metabolic (re)programming toward hepatic fat accumulation [5], allow suggesting the amelioration of chronic inflammation and long-term deterioration of liver function. Third, distinct interstitial macrophage populations coexist across tissues in specific subtissular niches where the absence of Lyve-1 macrophages exacerbates fibrosis [9].

5. Conclusions

In metabolic mouse models, the immunonutritional potential of *C. quinoa* enables a selective functional differentiation and function of myeloid cell population toward resolutive macrophage. This influence favors tissue integrity in conditions of caloric excess. These data warrant further human-based research.

Author Contributions: All authors contribute equally to this paper. All authors have read and agreed to the published version of the manuscript.

Funding: This research received no external funding.

Institutional Review Board Statement: Animal experiments were carried out in strict accordance with the recommendations in the Guide for the Care and Use of Laboratory Animals of CSIC (Consejo Superior de Investigaciones Científicas) and the protocol was approved by its Ethic Committee and the regional government (Ethic code, Proex 220/17).

Data Availability Statement: The data that support the findings of this study are available from the corresponding author upon reasonable request.

Acknowledgments: This work was financially supported by grants La ValSe-Food CYTED, (119RT0S 67) and Food4ImNut (PRP_PID-2019-107650RBC22, Ministry of Science, Innovation and Universities—MICIU).

Conflicts of Interest: The authors declare no conflict of interest.

References

1. Krenkel, O.; Hundertmark, J.; Abdallah, A.T.; Kohlhepp, M.; Puengel, T.; Roth, T.; Branco, D.P.P.; Mossanen, J.C.; Luedde, T.; Trautwein, C.; et al. Myeloid cells in liver and bone marrow acquire a functionally distinct inflammatory phenotype during obesity-related steatohepatitis. *Gut* **2020**, *69*, 551–563. [CrossRef] [PubMed]
2. Griffin, C.; Eter, L.; Lanzetta, N.; Abrishami, S.; Varghese, M.; McKernan, K.; Muir, L.; Lane, J.; Lumeng, C.N.; Singer, K. TLR4, TRIF, and MyD88 are essential for myelopoiesis and CD11c^{+} adipose tissue macrophage production in obese mice. *J. Biol. Chem.* **2018**, *293*, 8775–8786. [CrossRef] [PubMed]
3. Selma-Gracia, R.; Haros, C.M.; Laparra Llopis, J.M. Inclusion of *Salvia hispanica* L. and *Chenopodium quinoa* into bread formulations improves metabolic imbalances derived from a high-fat intake in hyperglycaemic mice. *Food Funct.* **2020**, *11*, 7994–8002. [CrossRef]

4. Srdić, M.; Ovčina, I.; Fotschki, B.; Haros, C.M.; Laparra, J.M. *C. quinoa* and *S. hispanica* L. seeds provide immunonutritional agonists to selectively polarize macrophages. *Cells* **2020**, *9*, 593. [CrossRef]
5. Chamorro-García, R.; Sahu, M.; Abbey, R.J.; Laude, J.; Pham, N.; Blumberg, B. Transgenerational inheritance of increased fat depot size, stem cell reprogramming, and hepatic steatosis elicited by prenatal exposure to the obesogen tributyltin in mice. *Environ. Health Perspect.* **2013**, *121*, 359–366. [CrossRef] [PubMed]
6. Yki-Järvinen, H. Non-alcoholic fatty liver disease as a cause and a consequence of metabolic syndrome. *Lancet Diabetes Endocrinol.* **2014**, *2*, 901–910. [CrossRef]
7. Kang, L.; Lantier, L.; Kennedy, A.; Bonner, J.S.; Mayes, W.H.; Bracy, D.P.; Bookbinder, L.H.; Hasty, A.H.; Thompson, C.B.; Wasserman, D.H. Hyaluronan accumulates with high-fat feeding and contributes to insulin resistance. *Diabetes* **2013**, *62*, 1888–1896. [CrossRef] [PubMed]
8. Li, B.; Guo, J.; Xi, Z.; Xu, J.; Zuo, Z.; Wang, C. Tributyltin in male mice disrupts glucose homeostasis as well as recovery after exposure: Mechanism analysis. *Arch. Toxicol.* **2017**, *91*, 3261–3269. [CrossRef] [PubMed]
9. Chakarov, S.; Lim, H.Y.; Tan, L.; Lim, S.Y.; See, P.; Lum, J.; Zhang, X.-M.; Foo, S.; Nakamizo, S.; Duan, D.; et al. Two distinct interstitial macrophage populations coexist across tissues in specific subtissular niches. *Science* **2019**, *363*, eaau0964. [CrossRef] [PubMed]

biology and life sciences forum

MDPI

Proceeding Paper

Andean Ancient Grains: Nutritional Value and Novel Uses †

Ritva Repo-Carrasco-Valencia *, Jaime Basilio-Atencio, Genny Isabel Luna-Mercado, Silvia Pilco-Quesada and Julio Vidaurre-Ruiz

CIINCA (Center of Innovation for Andean Grains), Universidad Nacional Agraria La Molina, Av. La Molina s/n, La Molina, Lima 12, Peru; Jaime.basilio@unas.edu.pe (J.B.-A.); gluna@unap.edu.pe (G.I.L.-M.); silviapilco@upeu.edu.pe (S.P.-Q.); vidaurrejm@lamolina.edu.pe (J.V.-R.)

* Correspondence: ritva@lamolina.edu.pe

† Presented at the III Conference la ValSe-Food and VI Symposium Chia-Link Network, Online, 15–17 November 2021.

Abstract: Quinoa, kañiwa, kiwicha and tarwi are ancient native crops from the Andes highlands of South America. Due to their remarkably high nutritional value, they offer major promise as ingredients in various food products. The aims of this study were to determine the nutritional value of certain varieties of quinoa, kañiwa, kiwicha and tarwi and to use these grains to develop novel, nutritious prototypes of products such as a malted beverage, extruded porridge, gluten-free bread and culinary dishes. The proximate, mineral and phenolic compound contents were evaluated in the Andean grains and final products. Two gluten-free breads were prepared, one made with quinoa and another made with kañiwa. An instant porridge prototype for child nutrition was developed. It had a protein content of 16% and it could, therefore, be considered to be a source of protein. The protein had a high in vitro digestibility (96.3%) and the chemical score was 0.92. The malted beverage prepared with quinoa and kiwicha had a protein content of 7.7%, which represents a value of 1.5 to 2 times more protein than dairy milk. The quinoa-amaranth beverage developed in this study is an excellent locally grown alternative to commercially available plant-based beverages usually made with soy, almond or oat, all of which are imported into Peru. Quinoa, kañiwa, kiwicha and tarwi are innovative, nutritious and tasty alternatives for restaurants seeking new ingredients for their recipes.

Keywords: *Amaranthus chenopodium*; extrusion; gluten-free; Lupinus; malted beverage

Citation: Repo-Carrasco-Valencia, R.; Basilio Atencio, J.; Luna Mercado, G.I.; Pilco-Quesada, S.; Vidaurre-Ruiz, J. Andean Ancient Grains: Nutritional Value and Novel Uses. *Biol. Life Sci. Forum* **2021**, *8*, 15. https://doi.org/10.3390/blsf2021008015

Academic Editors: Loreto Muñoz and Claudia M. Haros

Published: 17 January 2022

1. Introduction

The Andean region of South America is an important centre of domestication of food crops. This region has a great diversity of agroecological zones due to several climate and altitude differences (1500–4200 m). The use and cultivation of many of these plants decreased dramatically as a result of European colonization. The diets of Andean inhabitants changed and native grains were replaced by imported crops such as wheat, soy and rice. This change had long term effects and has affected the nutritional status in Peru and other Andean countries where malnutrition is common.

During the past decades, Andean crops have been "re-discovered" and attracted worldwide interest due to their remarkably high nutritional value and environmental adaptability. The most notable Andean grains are quinoa (*Chenopodium quinoa* Willd.), kañiwa (*Chenopodium pallidicaule* Aellen), kiwicha (*Amaranthus caudatus* L.) and tarwi (*Lupinus mutabilis*). These crops have adapted perfectly to the harsh environmental conditions of the Andes, being very resistant against drought, the salinity of the soil and frost. After the FAO declared 2013 as the "International Year of Quinoa", the cultivation and use of quinoa has extended beyond its area of origin [1].

Quinoa, kañiwa and kiwicha are pseudocereals whereas tarwi is a leguminous plant. The Andean pseudocereals have a relatively high protein content with an excellent biological quality [2,3]. Andean grains are gluten-free and could be used as ingredients in products for people suffering from coeliac disease [4]. Gluten-free products that are currently commercially available in Latin America lack the high nutritional value of Andean grains. In recent years, there has been a remarkable global interest in Peruvian haute cuisine, which has been ranked as one of the world's best. Since 2013, most Peruvian gourmet restaurants have included quinoa in their menu; however, other less-known crops are usually not included.

With the purpose of causing an impact on the Peruvian food sector, the aims of this study conducted under the Project Protein2Food were to determine the nutritional value of several varieties of quinoa, kañiwa, kiwicha and tarwi and to use these Andean native grains to develop novel and nutritive prototypes of products such as a malted beverage, extruded porridge, gluten-free bread and culinary dishes.

2. Materials and Methods

2.1. Materials

The following varieties of Andean grains were used to develop the prototypes: the Pasankalla and Chullpi varieties of quinoa and the Oscar Blanco and Centenario varieties of kiwicha, all of which were provided by "El Programa de Cereales y Granos Nativos de la Universidad Nacional Agraria La Molina". The variety of kañiwa was Illpa Inia from Puno. For tarwi, the variety Yunguyo was acquired from Agroinversiones Ogoríz S.R.L.-Cajamarca. For culinary dishes, the following quinoa varieties were used, all acquired from INIA: Blanca de Junin, Kancolla, Amarilla de Sacaca and CICA-18.

The following food products were developed: an extruded product for porridge, gluten-free bread and a malted beverage.

2.2. Methods

An analysis of the protein, carbohydrates, fat, ash, moisture and crude fibre was performed using the Official Methods of Analysis [5].

The total, soluble and insoluble dietary fibre were analysed by an enzymatic-gravimetric method according to the Official Method of the AOAC [5]. The total phenolic (TP) content was measured by spectrophotometry [6]. The chemical score of the protein for the selected pasta was evaluated based on the amino acid requirements for adults.

A specific volume of bread was measured by laser topography (BVM-6610, Perten Instruments, Sweden) and the specific volume (mL/g) was calculated by dividing the volume with the weight of the bread.

The potentiometric method of the AOAC [5] was used where the sample was diluted with distilled water (1:10) and the pH was measured using a pH meter (Hanna Instruments, HI2020). The acidity, in terms of succinic acid, was determined using a titrimetric method.

Statistical Analysis

All experimental analyses were performed in triplicate. The results were expressed as a mean ± standard deviation.

3. Results and Discussion

In Table 1, the chemical composition of the grains is presented. The protein content of the different quinoa varieties was 9.6–15.2%. In the case of kiwicha and kañiwa, this value was 14.3–15.4%. Tarwi had an extremely high protein content of 52%. Cordoba et al. [7] analysed the protein content of three tarwi varieties and they reported an average value of 54.4%, which is in accordance with our results.

The chemical composition of the final products can be seen in Table 1. The protein content for the gluten-free breads was 7.7% for the quinoa bread and 16.3% for the kañiwa bread. This high protein content for the kañiwa bread was due to the fact that this bread was made of 100% kañiwa flour without adding starch whereas the quinoa bread contained potato starch. The kañiwa bread could be considered to be a source of protein according to the nutrition claims of the European Union because more than 12% of the energy value of the food was provided by protein. It was possible to obtain a good quality bread with kañiwa without adding any starch (see pictures 1 and 2 for quinoa and kañiwa bread, respectively). This was interesting and indicated that kañiwa starch is different from quinoa starch and helps in the formation of gluten-free dough. Kañiwa has a high dietary fibre content that can help to strengthen the flour and has a positive influence on the final product. Kañiwa is reported to have a low amylose content; Cornejo and Rosell [8] reported that grain varieties with low amylose contents present low gelatinisation temperatures and soft gels, which is beneficial for baked products. Both breads could be considered to be high fibre because the content of this component was superior to 6 g/100 g of product. This is important because commercially available gluten-free products generally lack dietary fibre; they are made mainly of starch and white rice flours.

Table 1. Proximate compositions of Andean grains and products.

Grain Sample	Moisture g/100 g	Protein g/100 g	Fat g/100	Crude Fibre g/100	Ash g/100	Carbohydrates g/100	Ironm g/kg	Calcium mg/kg
Quinoa Pasankalla	8.03 ± 0.11	14.37 ± 0.06	5.77 ± 0.28	17.24 ± 0.04 *	2.34 ± 0.06	69.49	38.63 ± 3.35	1134.1 ± 44.5
Quinoa Chullpi	10.67 ± 0.31	9.64 ± 0.37	15.19 ± 1.09	6.22 ± 0.05	5.51 ± 0.37	64.58	n.d.	n.d.
Quinoa Blanca de Junin	10.46 ± 0.06	14.03 ± 0.19	7.88 ± 0.16	3.15 ± 0.13	2.68 ± 0.09	72.27	77.33 ± 5.508	1100 ± 10.3
Quinoa Kancolla	10.25 ± 0.08	14.41 ± 0.07	7.57 ± 0.23	3.31 ± 0.13	2.35 ± 0.04	72.37	83.00 ± 9.539	1100 ± 20.3
Quinoa Amarilla Sacaca	11.20 ± 0.14	14.10 ± 0.14	7.21 ± 0.73	3.82 ± 0.09	2.29 ± 0.16	72.58	86.33 ± 10.01	1300 ± 20.5
Quinoa CICA-18	10.75 ± 0.11	15.22 ± 0.13	7.23 ± 0.16	4.02 ± 0.02	2.48 ± 0.15	71.05	116.33 ± 24.02	1200 ± 30.3
Kañiwa	11.83 ± 0.07	14.68 ± 0.07	8.9 ± 0.11	13.69 ± 0.11 *	2.58 ± 0.08	69.49	104.4 ± 0.2	1528.1 ± 6.05
Kiwicha Oscar Blanco	6.91 ± 0.28	15.39 ± 0.19	13.71 ± 0.52	7.53 ± 0.39	3.80 ± 0.32	57.10	n.d.	n.d.
Kiwicha Centenario	10.25 ± 0.09	14.33 ± 0.05	7.15 ± 0.04	2.52 ± 0.05	2.58 ± 0.03	65.69	48.76 ± 3.20	937.30 ± 21.85
Tarwi Yunguyo	9.0 ± 0.06	52.42 ± 0.21	21.41 ± 0.02	7.47 ± 0.13	2.7 ± 0.02	14.48	162.00 ± 6.79	404.30 ± 8.49
Gluten-free quinoa bread	27.75 ± 0.1	7.74 ± 0.0	4.74 ± 0.0	14.38 ± 0.2 *	3.22 ± 0.0	84.08 ± 0.1	13.51 ± 1.75	453.20 ± 126.01
Gluten-free kañiwa bread	37.62 ± 2.15	16.29 ± 0.99	12.03 ± 0.64	13.16 ± 1.13 *	3.21 ± 0.18	64.94 ± 2.16	103.5 ± 0.28	1639.9 ± 7.66
Extruded porridge	7.1 ± 0.06	16 ± 0.30	7.28 ± 0.05	n.d.	2.6 ± 0.03	67.10	n.d.	n.d.
Malted beverage	90.02 ± 0.13	7.67 ± 0.11	0.28 ± 0.01	n.d.	0.04 ± 0.00	1.05 ± 0.00	n.d.	n.d.

* Total dietary fibre; n.d.: not determined.

The quality characteristics of the gluten-free quinoa and kañiwa bread are presented in Table 2. The specific volume of the kañiwa bread was 2.96 and 1.82 for the quinoa bread. This high specific volume could be due to the adequate hydration of the dough by the presence of components with a high water absorption capacity such as fibre, protein and lipids that help to maintain the viscosity and fluidity of the dough during fermentation and cooking (the bake loss was 40% and 30 % for kañiwa and quinoa, respectively.

Table 2. Quality characteristics of the optimized gluten-free breads based on quinoa and kañiwa.

Quality Parameters	Gluten-Free Bread	
	Kañiwa (100%)	Quinoa (44.5%)
Specific volume (mL/g)	2.96 ± 0.05	1.82 ± 0.10
Bake loss %	40.08 ± 1.25	30.46 ± 1.80
Crumb hardness (N)	0.77 ± 0.14	1.90 ± 0.30
Cohesiveness	0.42 ± 0.02	0.68 ± 0.00
Springiness	0.76 ± 0.03	0.85 ± 0.10
Gumminess (N)	0.34 ± 0.04	1.28 ± 0.20
Chewiness (N)	0.25 ± 0.03	1.09 ± 0.20

The protein content of the extruded porridge was 16%. This product could be considered to be a source of protein according to the nutrition claims of the European Union because more than 12% of the energy value of the food was provided by protein. The protein in the in vitro digestibility was 96.3%, which could be considered to be an appropriate value from a nutritional point of view. This value was superior to the value reported by Akande et al. [9] for amaranth-based mixtures. The chemical score was also high (0.92). This was because the Andean grains (quinoa, kiwicha and tarwi) complemented each other in their composition of essential amino acids.

Several culinary dishes were prepared using the different Andean grains in collaboration with chefs of Peruvian gourmet restaurants. In pictures 1–3, a few examples can be seen. These dishes presented excellent nutritional value, especially in their protein, bioactive compounds and dietary fibre content. The results of this research have been published in the cookery book "Andean Native Grains. Superfoods For The Kitchen" (Repo de Carrasco and Solorzano, 2020). It includes recipes of beverages, starters, main dishes and desserts as well as scientific information about the grains and prepared dishes. This book has had a very wide dissemination amongst people working with Peruvian gastronomy and food scientists. The recipes are currently used on the menus of Peruvian gourmet restaurants.

4. Conclusions

All varieties of Andean grains included to this study showed an excellent nutritional value with a high protein content; tarwi was especially rich in this nutrient. In addition, these grains are excellent sources of dietary fibre and phenolic compounds. Amongst them, kañiwa stands out. The oil content of tarwi is high and it could be an excellent source of edible oil. Regarding nutritionally important minerals, quinoa and kañiwa can be considered to be good sources of iron and calcium. These underutilized grains can be used as ingredients for a variety of food products such as gluten-free bread, plant-based beverages and children's food. In gluten-free products, Andean grains improve the nutritional and sensorial qualities.

In this study, two gluten-free formulations were developed, one based on quinoa and another based on kañiwa. Both breads had a particularly good nutritional composition, with high protein and dietary fibre contents and kañiwa was again outstanding in this aspect. Not only was the macronutrient content excellent in the kañiwa bread but also the micronutrient content excelled. Another interesting point is that this bread could be elaborated without adding any starch. It seems that kañiwa starch is different from quinoa starch and helps the formation of gluten-free dough.

The instant porridge based on extruded quinoa, kiwicha and tarwi flour can be considered to be a source of protein with excellent in vitro digestibility and a high chemical score. This product could be used for infant foods by local food industries and offered to governmental food aid programs in Peru in order to replace the use of imported ingredients such as wheat and soy.

The quinoa-amaranth beverage developed in this study offers an excellent locally grown alternative to commercially available plant-based beverages usually made by soy, almond or oat, which are all imported into Peru. Additionally, quinoa, kañiwa, kiwicha and tarwi are innovative, nutritious and tasty alternatives for restaurants seeking new ingredients for their recipes. This is the first time that these four crops have been included in the menu of Peruvian fine food restaurants.

Institutional Review Board Statement: Not applicable.

Informed Consent Statement: Not applicable.

Data Availability Statement: Not applicable.

Acknowledgments: This work was supported by grant Ia ValSe-Food-CYTED (119RT0567) and the Protein2Food project, which received funding from the European Union's Horizon 2020 Research and Innovation Framework Programme under grant agreement No 635727.

References

1. Jacobsen, S.E. The scope for adaptation of quinoa in Northern Latitudes of Europe. *J. Agron. Crop. Sci.* **2017**, *203*, 603–613. [CrossRef]
2. Mattila, P.; Mäkinen, S.; Eurola, M.; Jalava, T.; Pihlava, J.M.; Hellström, J.; Pihlanto, A. Nutritional Value of Commercial Protein-Rich Plant Products. *Plant Foods Hum. Nutr.* **2018**, *73*, 108–115. [CrossRef] [PubMed]
3. Repo-Carrasco-Valencia, R.; Vidaurre-Ruiz, J. Quinoa and Other Andean Ancient Grains: Supergrains for the Future. *Cereal Foods World* **2019**. Available online: https://www.cerealsgrains.org/publications/cfw/2019/September-October/Pages/CFW-64-5-0053.aspx (accessed on 25 November 2021).
4. Coda, R.; Rizzello, C.; Gobbetti, M. Use of sourdough fermentation and pseudo-cereals and leguminous flours for the making of functional bread enriched of γ-aminobutyric acid (GABA). *Int. J. Food Microbiol.* **2010**, *137*, 236–245. [CrossRef] [PubMed]
5. AOAC. *Official Methods of Analysis*, 17th ed.; Association of Official Analytical Chemists: Gaithersburg MD, USA; Washington, DC, USA, 2000.
6. Chávez, D.W.H.; Ascheri, J.L.R.; Carvalho, C.W.P.; Godoy, R.L.O.; Pacheco, S. Sorghum and roasted coffee blends as a novel extruded product: Bioactive compounds and antioxidant capacity. *J. Funct. Foods* **2017**, *29*, 93–103. [CrossRef]
7. Córdova-Ramos, J.; Glorio-Paulet, P.; Camarena, F.; Brandolini, A.; Hidalgo, A. Andean lupin (Lupinus mutabilis Sweet): Processing effects on chemical composition, heat damage and in vitro protein digestibility. *Cereal Chem.* **2020**, *97*, 827–835. [CrossRef]
8. Cornejo, F.; Rosell, C.M. Physicochemical properties of long rice grain varieties in relation to gluten free bread quality. *LWT-Food Sci. Technol.* **2015**, *62*, 1203–1210. [CrossRef]
9. Akande, O.A.; Nakimbugwe, D.; Mukisa, I.M. Optimization of extrusion conditions for the production of instant grain amaranth-based porridge flour. *Food Sci. Nutr.* **2017**, *5*, 1205–1214. [CrossRef] [PubMed]

MDPI
St. Alban-Anlage 66
4052 Basel
Switzerland
Tel. +41 61 683 77 34
Fax +41 61 302 89 18
www.mdpi.com

Biology and Life Sciences Forum Editorial Office
E-mail: blsf@mdpi.com
www.mdpi.com/journal/blsf